LECTURES ON PSEUDO-DIFFERENTIAL OPERATORS:

Regularity theorems and applications to non-elliptic problems

by

ALEXANDER NAGEL and E. M. STEIN

Princeton University Press

and

University of Tokyo Press

Princeton, New Jersey

1979

Published in Japan Exclusively
by University of Tokyo Press
in other parts of the world by
Princeton University Press

Printed in the United States of America
by Princeton University Press, Princeton, New Jersey

Library of Congress Cataloging in Publication Data will
be found on the last printed page of this book

Table of Contents

Preface

The theory of pseudo-differential operators (which originated as singular integral operators) was largely influenced by its application to function theory in one complex variable and regularity properties of solutions of elliptic partial differential equations. It is our goal here to give an exposition of some new classes of pseudo-differential operators relevant to several complex variables and certain non-elliptic problems. As such this monograph contains the details of the results we announced earlier in [34], together with some background material. What is presented below was the subject of a course given by the second author at Princeton University during the Spring term of 1978.

We are very happy to acknowledge the assistance given us by David Jerison. He prepared a draft of the lecture notes of the course and made several valuable suggestions which are incorporated in the text. We should also like to thank Miss Florence Armstrong for her excellent job of typing of the manuscript.

INTRODUCTION

The object of this monograph is to present the theory of certain new classes of pseudo-differential operators. These classes of operators are intended to satisfy two general requirements. On the one hand, they should be restrictive enough to be bounded in L^p, Lipschitz spaces, and Sobolev spaces. On the other hand, they should be large enough to allow for a description of the parametricies of some interesting non-elliptic differential and pseudo-differential operators and other operators such as the Cauchy-Szegö and Henkin-Ramirez integrals for strictly pseudo-convex domains.

Before discussing these new classes however, we will briefly recall some basic definitions, and outline the situation in the classical "elliptic" case.

A pseudo-differential operator, defined initially on the Schwartz class $\mathcal{S}(\mathbb{R}^n)$, has the form:

$$(1) \qquad f \longrightarrow a(x,D)(f(x)) = (2\pi)^{-n} \int_{\mathbb{R}^n} e^{i(x,\xi)} a(x,\xi) \hat{f}(\xi) d\xi$$

where $\hat{f}(\xi) = \int_{\mathbb{R}^n} e^{-i(x,\xi)} f(x)dx$ is the Fourier transform, and the symbol $a(x,\xi)$ is smooth, and has at most polynomial growth in $|\xi|$. One also imposes additional differential inequalities on this symbol. For example, the "classical elliptic symbols" of order m are defined by:

$$(2) \qquad S^m_{1,0} = \{a(x,\xi) \in C^\infty(\mathbb{R}^{2n}) \mid |\partial_x^\beta \partial_\xi^\alpha a(x,\xi)| \leq C_{\alpha\beta}(1+|\xi|)^{m-|\alpha|}\}.$$

One interest in these classes arises from the fact that a parametrix for an an elliptic differential operator of order m can be written as a pseudo-differential operator with symbol in $S^{-m}_{1,0}$.

The appearance of the Fourier transform in the definition (1) of a pseudo-differential operator often makes this the appropriate form for proving L^2 estimates. For example, if the symbol $a(x,\xi)$ is independent of x, the operator $a(x,D)$ is then just a Fourier multiplier operator, and the boundedness of $a(x,D)$ from $L^2(\mathbb{R}^n)$ to itself is equivalent to the uniform boundedness of the symbol $a(\xi)$. More generally, it is fairly easy to prove that if $a(x,\xi) \in S^0_{1,0}$ then $a(x,D)$ extends to a bounded operator from $L^2(\mathbb{R}^n)$ to itself.

However, if one wants to prove that pseudo-differential operators are bounded on L^p $(p \neq 2)$ or on Lipschitz spaces, one needs to represent these operators in another way as singular integral operators. These are operators of the form

$$f \longrightarrow Kf(x) = \int_{\mathbb{R}^n} K(x, x-z)\, f(z)\, dz . \tag{3}$$

Here the kernel $K(x,y)$ will in general be singular when $y = 0$, so the integral in (3) must be taken in a principal value sense. By explicitly writing out the Fourier transform in (1), and formally interchanging the order of integration, one sees that the kernel $K(x,y)$ of the operator is related to the symbol $a(x,\xi)$ by the formula:

$$K(x,y) = (2\pi)^{-n} \int_{\mathbb{R}^n} e^{i(y,\xi)} a(x,\xi)\, d\xi . \tag{4}$$

If $a(x,\xi) \in S^m_{1,0}$ is a classical elliptic symbol, one can use (4) to obtain estimates on the associated kernel, of the form

(5) $$|\partial_x^\alpha \partial_y^\beta K(x,y)| \leq C'_{\alpha\beta}|y|^{-n-m-|\beta|}$$

at least when $n+m+|\beta| > 0$. (For future reference, note that (5) is equivalent to

(5′) $$|\partial_x^\alpha \partial_y^\beta K(x,y)| \leq C''_{\alpha\beta}|y|^{-m-|\beta|}[\mathrm{vol}\,(B(x,|y|))]^{-1}$$

where $B(x,|y|)$ is the ball centered at x with radius $|y|$.)

Once one has estimates (5) or (5′), one can apply the classical Calderón-Zygmund theory to obtain L^p estimates $(1 < p < \infty)$ for the singular integral operator with kernel $K(x,y)$. For example, it follows easily from (5) or (5′) that one has the well-known condition:

(6) $$\int_{|x-x_0|>2|y-y_0|} |K(x,x-y) - K(x,x-y_0)|dx \leq C,$$

for a kernel $K(x,y)$ corresponding to a symbol $a(x,\xi) \in S^0_{1,0}$. It is this approach that we try to imitate in some non-elliptic situations.

In studying a variety of non-elliptic problems, several more general classes of pseudo-differential operators have been developed - for example by Hörmander, Calderón and Vaillancourt, Boutet de Monvel, Sjöstrand, Beals and Fefferman, and Beals. (See [1] and [25] for references.) In these classes, one again imposes certain conditions on the size of the derivatives of a symbol, but the conditions are different from those in (2). For example, one can allow a certain loss in $|\xi|$ with every x derivative, and a similar gain in $|\xi|$ with every ξ derivative. In all of these cases, it is proved, among other things, that symbols of order zero give rise to operators which are bounded on L^2. (The proofs, however, are considerably more delicate than in the classical elliptic case.) However, boundedness in L^p $(p \neq 2)$ or in Lipschitz spaces is in general false, and one does not obtain appropriate estimates for the associated singular integral operators.

On the other hand, Folland and Stein [16], and Rothschild and Stein [39] have shown that parametricies for certain hypoelliptic differential operators can be approximated, in an appropriate sense, by singular integral convolution operators on nilpotent Lie groups. Since a version of the Calderón-Zygmund theory is available in that context, they are able to prove sharp L^p and Lipschitz estimates for these parametricies. However, these operators were not realized as pseudo-differential operators.

We can now enunciate a basic guiding principle of our work: To treat only those classes of symbols for which one can prove that the corresponding operators have singular integral realizations with kernels having properties analogous to (5), (5′), and (6). In this sense our approach to pseudo-differential operators is essentially different from the generalizations which have been studied in the last dozen years. What the more general forms of (2), (5′), and (6) might be is not a simple matter, but it is in part motivated by the theory of singular integral operators on nilpotent groups, and the background for this is presented in Chapter 1.

Our theory then proceeds along the following lines:

1. A ρ function is introduced in the (x, ξ) space which reflects the

geometry of each particular situation and in terms of which we will control the size of symbols and their derivatives. This ρ function leads by duality to a basic family of "balls" in the x space. It is the pseudo-distance defined by these balls, and their volume in terms of which we estimate the kernels of our operators (when realized as (3)).

2. In this setup one can apply the Calderón-Zygmund theory (via a variant of (6)) to prove L^p estimates for our operators, assuming they are bounded in L^2. But here we must emphasize an important point. At this stage we work with a preliminary symbol class $\tilde{S}^m_\rho$ (defined very roughly by the requirement that the analogue of (5′) holds). But this class is too broad to allow L^2 estimates (in fact in the classical case it corresponds to $S^0_{1,1}$). So we must refine the class $\tilde{S}^m_\rho$; the resulting class, S^m_ρ, cannot be defined in terms of simple differential inequalities. The actual motivation for the definition we give is in terms of the explicit examples presented in Chapter 1. The class S^m_ρ also has the further property that it allows Lipschitz space and Sobolev space estimates - and this is carried out in Chapter 3. There are two types of estimates of this kind, isotropic and non-isotropic ones.

3. We then show that operators whose symbols belong to S^m_ρ arise in various applications such as:

(i) The Cauchy-Szegö integral and Henkin-Ramirez integral for strictly pseudo-convex domains.

(ii) The parametricies for $\Box_b$ on boundaries of strictly pseudo-convex domains, in the sub-elliptic case.

(iii) The parametricies of operators of Hörmander $X_0 + \sum_{j=1}^{m} X_j^2$, in the "step 2" case. (The higher step case needs a generalization of our theory; in this connection, see the announcement [35].)

(iv) The "oblique derivative" problem. Here a further extension of our symbol classes is needed, because in general the elliptic symbols do not belong to S_ρ^m, and the parametricies are a mixture of S_ρ^m symbols with elliptic symbols.

(v) The parametricies for the second-order singular operators of the type first studied by Kannai, e.g.,

$$\frac{\partial}{\partial x_0} \pm x_0 \frac{\partial^2}{\partial x_1^2} .$$

Chapter I. Homogeneous distributions

Chapter I may be thought of as a review of some known facts which are basic in motivating our theory. Proofs are for the most part only sketched.

§1. Homogeneity and dilations in $\mathbb{R}^n$

Denote $x = (x_1, \ldots, x_n)$. Fix positive exponents $a_1, \ldots, a_n$, and define $\delta_t(x) = (t^{a_1} x_1, \ldots, t^{a_n} x_n)$, $0 < t < \infty$. Observe that

(1) $\delta_s \circ \delta_t = \delta_{st}$

(2) $\delta_1 = \mathrm{Id}$

(3) $\delta_t(x) \to 0$ as $t \to 0$ for all x

(More generally, one might let δ_t be given by multiplication by the matrix $e^{A \log t}$ where A is a real matrix whose eigenvalues all have positive real part.)

The change of variable formula for dilations δ_t is given by

$$d(\delta_t x) = t^a dx$$

where $a = (a_1 + \ldots + a_n)$. (In general, $a = \operatorname{trace} A$.)

Denote Euclidean norm by $\|x\|$.

<u>Proposition 1.</u> <u>There exists a norm function</u> $|x|$ <u>satisfying</u>

(a) $|x| \geq 0$; <u>and</u> $|x| > 0$ <u>if and only if</u> $x \neq 0$

(b) $x \to |x|$ <u>is everywhere continuous and</u> C^∞ <u>on</u> $\mathbb{R}^n \setminus \{0\}$

(c) $|\delta_t x| = t|x|$ <u>for all</u> $t > 0$

(d) $|x| = 1$ iff $\|x\| = 1$. (This is just a normalization.)

Proof. Define $|x| = t$ if $\|\delta_{t^{-1}} x\| = 1$. To obtain smoothness, apply the implicit function theorem. The rest of the proposition is obvious.

Define polar coordinates by $x = (t, \sigma)$ where $t = |x|$ and $\sigma = \delta_{t^{-1}} x$ (i.e., $\|\sigma\| = |\sigma| = 1$).

Remark. $|x| \approx \Sigma |x_j|^{1/aj}$, since this holds when $|x| = 1$, and both sides transform the same way under the dilations δ_t.

Proposition 2. $dx = t^{a-1}\omega(\sigma)dt\,d\sigma$ where $d\sigma$ denotes the usual measure on the unit sphere and ω is a positive C^{∞} function on the sphere. The proof, a simple calculation of the Jacobian, is left to the reader. (See Fabes and Rivière [12], p.20.)

Corollary. There exists a constant $c > 0$ such that for all $f: (0,\infty) \to [0,\infty)$ measurable,

$$\int_{\mathbb{R}^n} f(|x|) \frac{dx}{|x|^a} = c \int_0^{\infty} f(r) \frac{dr}{r}$$

Remark. The corollary implies that $|x|^{\alpha}$ is locally integrable iff $\alpha > -a$ and $|x|^{\alpha}$ is integrable at infinity iff $\alpha < -a$.

Let $\lambda \in \mathbb{C}$. We say that f is homogeneous of degree λ if $f(\delta_t(x)) = t^{\lambda} f(x)$, $t > 0$. Let K be a distribution. (By distribution we shall always mean tempered distribution.) If K were a function homogeneous of degree λ, then $t^{\lambda}\int K(x)\varphi(x)dx = \int K(\delta_t x)\varphi(x)dx = \int K(x)\varphi(\delta_{t^{-1}} x)t^{-a}dx$. It is therefore natural to call a distribution K homogeneous of degree λ if

$$t^{-a}K(\varphi \circ \delta_{t^{-1}}) = t^{\lambda} K(\varphi) \quad \text{for every test function } \varphi.$$

A distribution K is C^∞ in an open set Ω if there is a function $f \in C^\infty(\Omega)$ such that $K(\varphi) = \int f\varphi$ for all $\varphi \in C_0^\infty(\Omega)$. We will call a distribution K of class λ if it is homogeneous of degree λ and C^∞ on $\mathbb{R}^n \backslash \{0\}$.

Theorem 1. K is a distribution of class λ if and only if $\hat{K}$ is a distribution of class $-a-\lambda$.

Proof. Because $\hat{K}(\varphi) = K(\hat{\varphi})$ and $(\varphi \circ \delta_{t^{-1}})^\wedge = t^a \hat{\varphi} \circ \delta_t$, it's easy to see that $\hat{K}$ is homogeneous of degree $-a-\lambda$.

Let K_0 denote the C^∞ function on $\mathbb{R}^n \backslash \{0\}$ that agrees with K there. Choose $\psi \in C_0^\infty$ such that $\psi \equiv 1$ in a neighborhood of 0.

$$K = \psi K + (1-\psi)K = \psi K + (1-\psi)K_0$$

ψK has compact support, so its Fourier transform is C^∞ (even analytic). Denote $K_1 = (1-\psi)K_0 = K_1$ is C^∞ everywhere and

$$(\Delta^M K_1)^\wedge = (4\pi^2 \|\xi\|^2)^M \hat{K}_1(\xi).$$

For sufficiently large M, using homogeneity, we see that $\Delta^M K_1$ decreases quickly enough at infinity so that $\Delta^M K_1 \in L^1$. Therefore, $(4\pi^2\|\xi\|^2)^M \hat{K}_1(\xi)$ is continuous, and $\hat{K}_1(\xi)$ is continuous except at the origin. Similarly, $x^\alpha K_1(x)$ is homogeneous (of degree $\lambda + \sum \alpha_j a_j$) for large x, so that any derivative of $\hat{K}_1$ is continuous outside the origin. qed.

Example 1. Suppose $\operatorname{Re}\lambda > -a$. Let $K(x)$ be a C^∞ function away from 0 that is homogeneous of degree λ; then $K \in L^1_{loc}$ (by Proposition 2). Therefore, K defines a distribution. Conversely, all distributions of class λ arise in this way. In fact, if K is such a distribution, let K_0 be as above

Then $K - K_0$ is a distribution of class λ supported at the origin. Thus $K - K_0$ is a sum of derivatives of the delta function at the origin. It is easy to check that $\left.\frac{\partial^\alpha}{\partial x^\alpha}\right|_{x=0}$ has homogeneity $-a - \sum a_j \alpha_j \leq -a$. Therefore $K - K_0 = 0$.

Example 2. Suppose $\lambda = -a$. In order that a function $K(x)$, C^∞ away from 0, of degree $-a$ define a distribution, it must have mean value zero on the sphere. Conversely, each distribution of class $-a$ is the sum of such a function and a constant multiple of the delta function at 0.

Proposition 3. More generally, suppose $K_0 \in C^\infty(\mathbb{R}^n \setminus \{0\})$, K_0 is homogeneous of degree λ. Then:

(a) There exists a distribution K that agrees with K_0 in $\mathbb{R}^n \setminus \{0\}$.

(b) K can be chosen to be of class λ if and only if $\int_{|x|=1} K_0(x) x^\alpha dx = 0$ for all multiindices α such that $\lambda = -a - \sum a_j \alpha_j$. (Notice that this condition is vacuous unless λ lies at certain exceptional points on the negative real axis $\lambda \leq -a$.)

(c) The distribution K of class λ is unique up to linear combinations of $\left.\frac{\partial^\alpha}{\partial x^\alpha}\right|_{x=0}$ for those α for which $\lambda = -a - \sum a_j \alpha_j$.

Proof sketch. Write $K_0(x) = \Omega(x)|x|^\lambda$, where $\Omega(x)$ is homogeneous of degree zero. Fix $\varphi \in \mathcal{S}$.

$$I_\lambda = \int_{\mathbb{R}^n} |x|^\lambda \, \Omega(x) \varphi(x) \, dx$$

converges absolutely for $\operatorname{Re} \lambda > -a$. It can be continued analytically to be meromorphic in $\mathbb{C}$. It has at most simple poles at the points $-a-\sum \alpha_j a_j$, and these poles vanish under the compatibility conditions of b). In fact, $\varphi(x) = \sum_{|\alpha| \leq N} c_\alpha x^\alpha + R_N(x)$. Let $I'_\lambda = \int_{|x|\leq 1} |x|^\lambda \Omega(x)\varphi(x)\,dx$. Clearly $I_\lambda - I'_\lambda$ is entire. The main part of I'_λ is (by analytic continuation)

$$\sum C_\alpha \int_{|x|\leq 1} \Omega(x)\, x^\alpha |x|^\lambda dx = \sum C'_\alpha \int_0^1 r^{\lambda+\sum \alpha_j a_j + a - 1} dr$$

$$= \sum \frac{C'_\alpha}{\lambda + \sum \alpha_j a_j + a}.$$

Thus poles arise only when $\lambda = -a - \sum \alpha_j a_j$, and in that case the pole vanishes if we impose the condition(s) $C'_\alpha = 0$, which are equivalent to those stated in b).

We carry out the argument in more detail for the case $\lambda = -a$.

<u>Proposition 3'.</u> <u>Suppose</u> $K_0(x)$ <u>is homogeneous of degree</u> $-a$ <u>and</u> C^∞ <u>away from</u> 0. $K_0(x)$ <u>extends to a distribution of class</u> $-a$ <u>if and only if</u> $\int_{|x|=1} K_0(x)\,dx = 0$.

<u>Proof.</u> Suppose K_0 has mean value zero. We give an alternative definition: We will define $K(\varphi) = \lim_{\varepsilon\to 0} \int_{|x|\geq \varepsilon} K_0(x)\varphi(x)\,dx$. Note that there exists $b > 0$ such that $\|x\| \leq c|x|^b$ for $\|x\| < 1$. (In fact, $b = 1/\max a_j$ works because $|x| \simeq \sum |x_j|^{1/a_j}$.) It follows that

$$\int_{\varepsilon<|x|} K_0(x)\varphi(x)dx = \int_{\varepsilon<|x|<1} K_0(x)(\varphi(x)-\varphi(0))dx + \int_{|x|>1} K_0(x)\varphi(x)dx$$

converges absolutely as $\epsilon \to 0$. Thus $K(\varphi)$ is well defined. Its homogeneity is also obvious.

Now suppose K_0 does not have mean value zero. There is a constant $c \neq 0$ such that

$$K_0(x) = K_1(x) + c|x|^{-a} \qquad \text{where } K_1(x)$$

is homogeneous of degree $-a$ and has mean value zero on the sphere. Since K_1 defines a distribution of class $-a$ we are reduced to showing that $c|x|^{-a}$ does not. Denote

$$\tilde{K}(\varphi) = c\left\{\int_{|x| \leq 1} \left(\frac{\varphi(x) - \varphi(0)}{|x|^a}\right) dx + \int_{|x| > 1} \frac{\varphi(x)}{|x|^a}\, dx\right\}. \tag{4}$$

$\tilde{K}$ is a distribution and agrees with $c|x|^{-a}$ away from 0. Assume there exists a distribution K of class $-a$ that agrees with $c|x|^{-a}$ away from 0. Then $(K - \tilde{K})(\varphi) = \sum c_\alpha \frac{\partial^\alpha \varphi}{\partial x^\alpha}(0)$. Choose φ so that $\varphi(0) \neq 0$, but $\frac{\partial^\alpha \varphi}{\partial x^\alpha}(0) = 0$ for all other α that occur in the sum. The formula implies

$$(K - \tilde{K})(\varphi \circ \delta_t) = c_0 \varphi(0) = (K - \tilde{K})(\varphi) \qquad \text{for all } t > 0.$$

By homogeneity, we also have $K(\varphi \circ \delta_t) = K(\varphi)$ for all $t > 0$. Therefore $\tilde{K}(\varphi \circ \delta_t) = \tilde{K}(\varphi)$ for all $t > 0$. But a change of variable in (4) shows that

$$\tilde{K}(\varphi \circ \delta_t) - \tilde{K}(\varphi) = -c\varphi(0) \int_{1 \leq |x| \leq t} \frac{dx}{|x|^a} = -c\, c'\varphi(0) \log t \neq 0,$$

a contradiction.

Remark. Let K_0 be homogeneous of degree $-a$. We have shown implicity that if K_0 has mean value zero with respect to one homogeneous norm $| \ |_1$, then it has mean value zero with respect to any homogeneous norm $| \ |_2$ (satisfying Proposition 1 a), b), c) but not necessarily the normalization d)). In fact, we can see this directly by observing that

$$\left\{ \int_{1 \leq |x|_1 \leq b} K_0(x)dx - \int_{1 \leq |x|_2 \leq b} K_0(x)dx \right\} = - \int_{1 \leq |x|_2 \leq b} K_0(x)dx$$

The left-hand side is bounded, but the right-hand side would grow like $-\log b$ if K_0 did not have mean value zero with respect to $| \ |_2$.

Examples. (Recall the convention $\hat{f}(\xi) = \int e^{2\pi i x \xi} f(x)dx$.)

1. (See Stein [41], Chapter III.) Let $a_1 = a_2 = \ldots = a_n = 1$.

$$(|x|^{-n+\alpha})^\wedge = \gamma_\alpha |\xi|^{-\alpha} \qquad \text{where } \gamma_\alpha = \frac{\pi^{n/2 - \alpha}\Gamma(\alpha/2)}{\Gamma(n/2 - \alpha/2)}$$

when $\alpha \neq 0, -2, -4, \ldots, \neq n, n+2, n+4, \ldots$, taken in the sense of Proposition 3 b).

More generally, if P_k is a harmonic polynomial on $\mathbb{R}^n$, homogeneous of degree k, then

$$\left(\frac{P_k(x)}{|x|^k} |x|^{-n+\alpha}\right)^\wedge = \gamma_{k,\alpha} \frac{P_k(\xi)}{|\xi|^{k+\alpha}} \qquad \text{where}$$

$$\gamma_{k,\alpha} = i^k \pi^{n/2-\alpha}\, \Gamma\!\left(\frac{k}{2} + \frac{\alpha}{2}\right) \Big/ \Gamma\!\left(\frac{k}{2} + \frac{n}{2} - \frac{\alpha}{2}\right)$$

when $\alpha \neq -k, -k-2, \ldots, \neq k+n, k+n+2, \ldots$.*

*For the other values of α these identities can be suitably interpreted if one takes into account the poles and zeroes of $\gamma_{k,\alpha}$.

2. $a_1 = a_2 = \ldots = a_n = 1, \quad a_{n+1} = 2.$

$$K(x,t) = \begin{cases} (4\pi)^{n/2} t^{-n/2} e^{-|x|^2/4t} & t > 0 \\ 0 & t \leq 0 \end{cases}$$

with $x = (x_1, \ldots, x_n), \quad t = x_{n+1}$.

$$\hat{K}(\xi) = \left(4\pi^2 \sum_{j=1}^{n} \xi_j^2 - 2\pi i \xi_{n+1}\right)^{-1}$$

Thus K is the fundamental solution to the heat equation

$$\left(\frac{\partial}{\partial t} - \Delta_x\right) K(x,t) = \delta$$

A fundamental fact connecting homogeneous distributions and operators is the following:

Theorem 2. Suppose K is homogeneous of degree $-a$ (or more generally of degree λ with $\operatorname{Re}\lambda = -a$) and is C^∞ away from the origin. Then $Tf = K * f$ is a bounded operator on L^2.

The Proof of Theorem 2 is immediate from the fact that $|\hat{K}|$ is homogeneous of degree 0, and hence bounded.

Remark. We will see later that T extends to a bounded operator on L^p, $1 < p < \infty$.*

§2. Homogeneous groups

We shall now sketch the background for an important generalization of Theorem 2. Consider $\mathbb{R}^n$ with a group multiplication that makes it a Lie group. Assume (for simplicity) that $x_1, \ldots, x_n$ are canonical

*See also Fabes and Rivière [12].

coordinates. (This means that they are the coordinates given by the exponential map. In particular, $x^{-1} = -x$ and a straight line through the origin is a one-parameter subgroup.) Suppose we have a one-parameter set of dilations $\delta_t(x) = (t^{a_1} x_1, \ldots, t^{a_n} x_n)$ that are automorphisms of the group. We will call a group with dilations a homogeneous group.

This situation arises often. If G is a semi-simple Lie group then G = KAN, where K is a maximal compact subgroup, A abelian, and N nilpotent. (This is the Iwasawa decomposition of G.) A acts on N by dilations, and $L^2(N)$ is of interest in representation theory. For example, let G = SU(n, 1), the biholomorphic self-mappings of the unit ball B in $\mathbb{C}^n$. Then K is the stability group of the origin and $B \simeq G/K$. N is the Heisenberg group and can be identified with the boundary of (the unbounded realization of) B. The abelian group A is one-dimensional and acts on N as do our dilations. We will return to this example in the next section.

We add two remarks. First, all homogeneous groups are nilpotent. This is because the Lie algebra can be split into a sum of sub-spaces $\mathfrak{A}_{a_j}$ of vector fields X satisfying

$$X(f \circ \delta_t) = t^{a_j} (Xf) \circ \delta_t .$$

Because $[\mathfrak{A}_{a_j}, \mathfrak{A}_{a_k}] \subset \mathfrak{A}_{a_j + a_k}$ and because there are only finitely many indices a_j, $\mathfrak{A}$ is nilpotent. Second, dx (Lebesgue measure on $\mathbb{R}^n$) is bivariant Haar measure. This is contained in a general theorem about the measure on the group inherited from the Lie algebra via the exponential mapping.* In our case a simpler proof can be given. Suppose that

*See Helgason [21].

$a_1 \leq a_2 \leq \ldots \leq a_n$. If $z = xy$, then

$$z_k = x_k + y_k + R_k(x,y), \qquad \text{where}$$

R_k is a polynomial in x and y with terms of (usual) degree at least two. But clearly

$$R_k(\delta_t x, \delta_t y) = t^{a_k} R_k(x,y)$$

so that R_k depends only on $x_1, \ldots, x_{k-1}, y_1, \ldots, y_{k-1}$. Thus the Jacobian matrix for $x \to xy$ (and similarly for $x \to yx$) is lower triangular with 1 in each entry on the diagonal.

Let f be a function on a homogeneous group H. $L_y f(x) = f(yx)$ and $R_y f(x) = f(xy)$ define left and right translation, respectively. An operator T is left invariant if $L_y(Tf) = T(L_y f)$, and similarly for right invariance. For example, right translation is left invariant. Define convolution by $f * g(x) = \int_H f(xy^{-1})g(y)dy = \int_H f(y)g(y^{-1}x)dy$. Since $f \to f * g$ is just a "sum" of right translations, it is reasonable to expect (and easy to show) that it is left invariant. In general, it is helpful to think of any left invariant operator as being given (at least formally) by

$$f \to f * K, \text{ where } K \text{ is some distribution.}$$

Let K be a distribution which is a C^∞ function away from the origin and is homogeneous of degree λ, $\operatorname{Re}\lambda = -a$. Define $Tf = f * K$.

Theorem 2′. T extends to a bounded operator on L^2.

Proof. We need the following lemma

Lemma. Suppose $\{T_j\}$ is a family of bounded operators on Hilbert space,

$a(j) \geq 0$ <u>and</u> $A = \sum_{j=-\infty}^{\infty} a(j) < \infty$. <u>Assume that</u>

(i) $\quad \|T_j T_k^*\| \leq a(j-k)^2$

(ii) $\quad \|T_j^* T_k\| \leq a(j-k)^2$.

<u>Then</u> $\|\sum T_j\| \leq A$.

<u>Proof of the lemma.</u> Suppose that there are N operators. Denote $T = \sum_{j=1}^{N} T_j$. $T\,T^*$ is self-adjoint, so we can compute its spectral norm:

$$\|T\|^2 = \|T\,T^*\| = \|(T\,T^*)^n\|^{1/n}$$

$(T\,T^*)^n = \sum T_{i_1} T_{i_2}^* T_{i_3} T_{i_4}^* \ldots T_{i_{2n-1}} T_{i_{2n}}^*$ where each i_k runs from 1 to N.

We can estimate each term of the sum in two ways. First,

$$\|T_{i_1} \ldots T_{i_{2n}}^*\| \leq \|T_{i_1} T_{i_2}^*\| \ldots \|T_{i_{2n-1}} T_{i_{2n}}^*\|$$

$$\leq a(i_1 - i_2)^2 \ldots a(i_{2n-1} - i_{2n})^2$$

Second,

$$\|T_{i_1} \ldots T_{i_{2n}}^*\| \leq \|T_{i_1}\| \, \|T_{i_2}^* T_{i_3}\| \ldots \|T_{i_{2n-2}}^* T_{i_{2n-1}}\| \, \|T_{i_{2n}}^*\|$$

$$\leq A^2 a(i_2 - i_3)^2 a(i_4 - i_5)^2 \ldots a(i_{2n-2} - i_{2n-1})^2$$

(Note that $\|T_{i_1}\| \leq A$ because $\|T_{i_1}\| = \|T_{i_1} T_{i_1}^*\|^{\frac{1}{2}} \leq a(0) \leq A$, and similarly for $\|T_{i_{2n}}^*\|$.)

Now use the geometric mean of these two estimates.

$$\|(TT^*)^n\| \leq \sum_{i_1,\ldots,i_{2n}} A\, a(i_1 - i_2)\, a(i_2 - i_3) \ldots a(i_{2n-1} - i_{2n}).$$

If we sum first over i_{2n}, we get

$$\sum_{i_1,\ldots,i_{2n-1}} A^2 a(i_1 - i_2)\, a(i_2 - i_3) \ldots a(i_{2n-2} - i_{2n-1}).$$

If we continue summing over i_{2n-1}, i_{2n-2}, etc. we end up with

$$\Sigma_{i_1} A^{2n} = N A^{2n}.$$

Therefore, $\|TT^*\| \leq N^{1/n} A^2$, and if we take the limit as $n \longrightarrow \infty$, the lemma is proved.

We now pass to the proof of Theorem 2′. For simplicity assume that $\lambda = -a$. Without loss of generality we may assume that the distribution K is given by a C^∞ function $K(x)$ with mean value zero. We will decompose $K(x)$ into a sum of terms $K_j(x)$ with support in the annulus $2^j < |x| \leq 2^{j+1}$, roughly speaking. In order to do so in a smooth way choose $\varphi_0 \in C^\infty[0,\infty)$ such that

$$\varphi_0(t) = \begin{cases} 1 & 0 \leq t \leq 1 \\ 0 & t \geq 2 \end{cases}.$$

Define $\varphi(x) = \varphi_0(|x|)$ and let

$$K_j(x) = (\varphi(2^{-j}x) - \varphi(2^{-j+1}x))\, K(x).$$

Because $\Sigma_j(\varphi(2^{-j}x) - \varphi(2^{-j+1}x)) \equiv 1$ for all $x \neq 0$, it follows that $K(x) = \sum_j K_j(x)$. Denote $T_j f = f * K_j$.

$$\|T_j\| = \|K_j\|_{L^1} \leq \text{const} \int_{2^{j-1}}^{2^{j+2}} \frac{dr}{r} \leq C^{\dagger}, \qquad \text{independent of } j$$

The adjoint of T_j, T_j^*, is given by $T_j^* f = f * K_j^*$ where $K_j^*(x) = \overline{K_j(x^{-1})}$. Recalling that $|x| = |x^{-1}|$, we see easily that K_j^* has the same properties as K_j, and so can be treated similarly.

In order to apply the lemma we shall show that

$$\text{(5)} \qquad \|T_j^* T_\ell\| \leq C\, 2^{-b|j-\ell|}$$

$$\text{(6)} \qquad \|T_j T_\ell^*\| \leq C\, 2^{-b|j-\ell|}$$

where $b > 0$ is the exponent for which $\|x\| \leq C|x|^b$ for $\|x\| \leq 1$ (see the proof of Proposition 3′).

Case a) $\ell > j + c_0$, where c_0 is some large constant.

$\|T_j^* T_\ell\| \leq \|K_\ell * K_j^*\|_{L^1}$ and

$$\int K_\ell(xy^{-1}) K_j^*(y)dy = \int (K_\ell(xy^{-1}) - K_\ell(x)) K_j^*(y)dy$$

The integrand vanishes unless $|xy^{-1}| \approx 2^\ell$ and $|y| \approx 2^j$. The weak triangle law implies that for sufficiently large c_0 we also have $|x| \approx 2^\ell$. Assume for the moment that we have proved the estimate

$$\text{(7)} \qquad |K_\ell(xy^{-1}) - K_\ell(x)| \leq C|y|^b / |x|^{a+b} \qquad \text{for } |x| \geq C|y|$$

Then

---

†By the corollary of Proposition 2.

$$\int |K_\ell(xy^{-1}) - K_\ell(x)|\ |K_j^*(y)|dy \leqq C \int \frac{|y|^b}{|x|^{a+b}} |K_j^*(y)|dy$$

$$\leqq C\, 2^{-(a+b)\ell} \int_{|y|\approx 2^j} |y|^b \frac{dy}{|y|^a} \leqq C\, 2^{-(a+b)\ell+bj} .$$

$K_\ell * K_j^*(x)$ is supported in the set $|x| \leq C 2^\ell$, whose total measure is $C 2^{a\ell}$. Therefore,

$$\|K_\ell * K_j^*\|_{L^1} \leqq C 2^{a\ell} 2^{-(a+b)\ell+bj} \leqq C 2^{-b(\ell-j)} .$$

Let us return to the proof of (7). A dilation and application of the formula $K_\ell(x) = 2^{a\ell} K_0(2^{-\ell}x)$ reduces (7) to

$$|K_0(xy^{-1}) - K_0(x)| \leq C|y|^b / |x|^{a+b}, \quad \text{for } |x| \geq C|y|$$

Now $|x| \approx 1$ on the support of $K_0(x)$ and the support of $K_0(xy^{-1})$ as a function of x, provided $|x| \geq C|y|$. Since K_0 is smooth,

$$|K_0(xy^{-1}) - K_0(x)| \leqq C\|y\| \leqq C|y|^b \leqq C|y|^b / |x|^{a+b}$$

This concludes the proof of (5). (6) is proved in a similar way.

<u>Case b)</u> $|j-\ell| \leq c_0$. Here we make a trivial estimate $\|T_\ell T_j^*\| \leqq \|T_\ell\|\ \|T_j^*\| \leqq$ const.

<u>Case c)</u> $j > \ell + c_0$. This case is similar to case a) qed.

<u>Theorem 3.</u> T <u>extends to a bounded operator on</u> L^p, $1 < p < \infty$.

The idea of the proof will be subsumed by that of a more general theorem proved in the next chapter.

§ 3. Homogeneous distributions on the Heisenberg group

3.1 The unit ball $B \subset \mathbb{C}^{n+1}$ given by $B = \{w \in \mathbb{C}^{n+1}: |w_1|^2 + \ldots + |w_{n+1}|^2 < 1\}$ is biholomorphic to the unbounded "half-space", defined by

$$\mathfrak{D} = \{z: \operatorname{Im} z_{n+1} > |z_1|^2 + \ldots + |z_n|^2\},$$

under the transformations

$$w_{n+1} = i\left(\frac{1-z_{n+1}}{1+z_{n+1}}\right); \; w_k = \frac{z_k}{1+z_{n+1}}, \quad k=1,\ldots,n$$

$$z_{n+1} = \frac{i-w_{n+1}}{i+w_{n+1}}; \; z_k = \frac{2iw_k}{i+w_{n+1}}, \quad k=1,\ldots,n$$

Denote by $\rho(z) = \operatorname{Im} z_{n+1} - \sum_{j=1}^{n} |z_j|^2$, the "height" function on $\mathfrak{D}$. Notice that $\mathfrak{D} = \{z: \rho(z) > 0\}$.

The Heisenberg group is the space $\mathbb{H}^n = \{(\zeta, t): \zeta \in \mathbb{C}^n, t \in \mathbb{R}\} \simeq \mathbb{C}^n \times \mathbb{R}$, with group law $(\zeta, t)(\zeta', t') = (\zeta + \zeta', t + t' + 2\operatorname{Im} \zeta \cdot \bar{\zeta}')$, $\zeta \cdot \bar{\zeta}' = \sum_{j=1}^{n} \zeta_j \cdot \bar{\zeta}_j'$.

The group $\mathbb{H}^n$ can be identified with certain (biholomorphic) self-mappings of $\mathfrak{D}$ that preserve ρ, as follows. Suppose $h = (\zeta, t)$, then

$$h: (z', z_{n+1}) \to (z' + \zeta, z_{n+1} + t + i|\zeta|^2 + 2iz' \cdot \bar{\zeta})$$

The reader may check that $h_1(h_2(z)) = (h_1 h_2)(z)$ and $\rho(h(z)) = \rho(z)$. Also, the identification $h \longleftrightarrow (\zeta, t + i|\zeta|^2) = h(0)$ gives a one-to-one correspondence between $\mathbb{H}^n$ and $b\mathfrak{D}$. We will denote by $d\beta$ the measure on $b\mathfrak{D}$ identified with bivariant Haar measure (= Euclidean measure $d\zeta\, dt$) on $\mathbb{H}^n$.

The left action of $\mathbb{H}^n$ on $\mathfrak{D}$ is analogous to the action by translation of $\mathbb{R}^n$ on the upper half space $\mathbb{R}^{n+1}_+$. Corresponding to rotations are the unitary transformations $u \in U(u)$, that act on $\mathcal{D}$ by $u: (z', z_{n+1}) \to (u(z'), z_{n+1})$. (Note that $\rho(u(z)) = \rho(z)$.) The rotations also act on $\mathbb{H}^n$: $(\zeta, t) \to (u(\zeta), t)$. Finally, the dilations

$$\delta(z) = (\delta z_1, \ldots, \delta z_n, \delta^2 z_{n+1}), \quad \delta > 0$$

also preserve $\mathcal{D}$, as we see from the relation

$$\rho(\delta z) = \delta^2 \rho(z).$$

The same dilations act on $\mathbb{H}^n$ $(\leftrightarrow b\mathcal{D})$

$$\delta(\zeta, t) = (\delta\zeta, \delta^2 t).$$

Thus $\mathbb{H}^n$ is a homogeneous group with homogeneous degree $a = 2n+2$.

3.2 The Cauchy-Szegö kernel

We want to find the kernel S satisfying

(8) $S(z,w)$ is defined on $\mathcal{D} \times \mathcal{D} \setminus$ diagonal of $b\mathcal{D}$.

(9) $S(z,w) = \overline{S(w,z)}$.

(10) For each w, $S(z,w)$ is a holomorphic function of z in $\mathcal{D}$.

(11) If f is holomorphic in $\mathcal{D}$ and "nice" near $b\mathcal{D}$, then

$$f(z) = \int_{b\mathcal{D}} S(z,w) f(w)\, d\beta(w).$$

Assuming that there exists a unique kernel S of this kind, we will be able to discover its form using the action of $\mathbb{H}^n$, $U(n)$, and dilations on $\mathcal{D}$.

Suppose $h \in \mathbb{H}^n$, then since $f(h(z))$ is holomorphic,

$$f(h(z)) = \int_{b\mathcal{D}} S(z,w)\, f(h(w))\, d\beta(w).$$

A change of variable implies

$$f(z) = \int_{b\mathcal{D}} S(h^{-1}z, h^{-1}w)\, f(w)\, d\beta(w) \qquad \text{for all } h \in \mathbb{H}^n.$$

By uniqueness, $S(z,w) = S(h^{-1}z, h^{-1}w)$ for all $h \in \mathbb{H}^n$. Suppose $w \in b\mathcal{D}$ and $z \in \mathcal{D}$. Choose $h_w \in \mathbb{H}^n$ so that $h_w(0) = w$. Then

(12) $$S(z,w) = S(h_w^{-1} z, 0)$$

We are now reduced to finding $S(z,0)$. Similar arguments show that

(13) $$S(u(z),0) = S(z,0) \qquad \text{for all } u \in U(n)$$

(14) $$S(\delta z, 0) = \delta^{-2n-2} S(z,0) \qquad \text{for all } \delta > 0$$

It follows from (10) and (13) that $S(z,0)$ depends only on z_{n+1}. Therefore, (14) implies that

$$S(z,0) = c_n z_{n+1}^{-n-1}$$

Now, with the help of (12), we can write down a formula for S, namely,

(15) $$S(z,w) = c_n\left(i(\overline{w}_{n+1} - z_{n+1}) - 2\sum_{k=1}^{n} z_k \overline{w}_k\right)^{-n-1}$$

<u>Theorem 4.</u> <u>The function</u> S <u>given by</u> (15) <u>with</u> $c_n = 2^{n-1} n!/\pi^{n+1}$ <u>satisfies</u> (8) - (11).

Before we begin the proof we would like to express S in a somewhat different form. For each $z \in \mathcal{D}$, we can write $z = z_b + \underline{i}\rho$ where $\underline{i} = (0,0,\ldots,0,i)$, $\rho = \rho(z)$, and $z_b \in b\mathcal{D}$.

$$S(z,w) = S(h_w^{-1}(z), 0) = S(h_w^{-1} h_{z_b}(0) + \underline{i}\rho, 0)$$

For each $\rho > 0$ define S_ρ on $\mathbb{H}^n$ by

$$S_\rho(h) = S(h(0) + \underline{i}\rho, 0) .$$

With this notation,

$$f(z) = f(z_b + \underline{i}\rho) = \int_{\mathbb{H}^n} S_\rho(h^{-1} h_{z_b}) f(h(0)) dh$$

Thus the Cauchy-Szegö kernel can be viewed as a convolution kernel on $\mathbb{H}^n$ on each level set of ρ.

And it is easy to check that

$$(16) \qquad S_\rho(\zeta, t) = c_n(|\zeta|^2 + \rho - it)^{-n-1} .$$

Proof of Theorem 4. Properties (8), (9), and (10) are obvious. We need only check (11). The case $n = 0$ is the usual Cauchy formula

$$(17) \qquad f(z) = \frac{1}{2\pi i}\int_{-\infty}^{\infty} \frac{f(w)}{(\overline{w} - z)} d\beta(w), \quad \operatorname{Im} z > 0,$$

for $f(z)$ holomorphic for $\operatorname{Im} z > 0$ and sufficiently small at infinity. A simple consequence is

$$(18) \qquad f(z) = \frac{n(-2i)^n}{2\pi i} \iint_{\operatorname{Im} w > 0} \frac{f(w)(\operatorname{Im} w)^{n-1}}{(\overline{w} - z)^{n+1}} dw, \quad n \geq 1,$$

where dw denotes Lebesgue measure in the w-plane. To prove (18), rewrite (17) as

$$f(z + 2iy) = \frac{1}{2\pi i}\int_{-\infty}^{\infty} \frac{f(s+iy)}{(s - iy - z)} ds, \quad \operatorname{Im} z > 0, \ y > 0,$$

and apply the formula $\psi(0) = \frac{(-1)^n}{(n-1)!} \int_0^\infty \left[\frac{d^n}{dy^n}\psi(y)\right] y^{n-1}\, dy$ to the function $\psi(y) = f(z+2iy)$.

To prove (11) in general, we need only do so for one value of $z \in \mathfrak{D}$, because we can translate, rotate, and dilate that value to obtain any other. We choose $z = \underline{i} = (0, \ldots, 0, i)$. We must show that

$$f(\underline{i}) = c_n \int \frac{f(w_1, \ldots, w_{n+1})}{(i(\overline{w}_{n+1} - i))^{n+1}}\, d\beta(w)$$

The right-hand side can be written as an integral on the Heisenberg group with $\zeta = (w_1, \ldots, w_n)$ and $w_{n+1} = t + i|\zeta|^2$ as

$$c_n \iint \frac{f(\zeta, t+i|\zeta|^2)}{(i(t-i|\zeta|^2 - i))^{n+1}}\, d\zeta\, dt\,.$$

Change to polar coordinates ($r = |\zeta|$, $\sigma_n = 2\pi^n/(n-1)! =$ vol. of the unit $2n-1$ sphere) to obtain

$$c_n \sigma_n \iint \int_{|\zeta|=r} \frac{f(\zeta, t+ir^2)}{(i(t-ir^2-i))^{n+1}}\, r^{2n-1}\, d\sigma(\zeta)\, dr\, dt\,.$$

Next apply the mean-value theorem, change coordinates to $w = t+ir^2$, and use (18):

$$c_n \sigma_n \iint \frac{f(0, t+ir^2)}{(i(t-ir^2-1))^{n+1}}\, r^{2n-1}\, dr\, dt$$

$$= \frac{c_n \sigma_n}{2} \iint_{\operatorname{Im} w > 0} \frac{f(0,w)}{(i(\overline{w}-i))^{n+1}}\, (\operatorname{Im} w)^{n-1}\, dw$$

$$= f(\underline{i}).$$

qed

We wish to generalize the classical Plemel formula. For $f \in C_0^\infty(\mathbb{R})$, $z = x + iy$,

$$\lim_{y \to 0} \frac{1}{2\pi i} \int_{-\infty}^{\infty} \frac{f(t)}{t-z}\, dt = \frac{1}{2} f(x) + \text{P.v.}\ \frac{1}{2\pi i} \int_{-\infty}^{\infty} \frac{f(t)}{t-x}\, dt$$

Recall that $S_\rho(\zeta, t) = c_n(|\zeta|^2 + \rho - it)^{-n-1}$. This makes sense even when $\rho = 0$. Furthermore, S_0 is homogeneous of degree $-2n-2 = -a$.

Theorem 5. $S_0(\zeta, t)$ has mean-value zero (so that p.v. (S_0) is defined as a distribution) and $S_\rho \to \frac{1}{2}\delta + \text{p.v.}(S_0)$ as a distribution.

Proof. One can compute that S_0 has mean-value zero directly, but we can avoid the computation as follows:

$$\lim_{\rho \to 0} \frac{c_n}{in}(|\zeta|^2 + \rho - it)^{-n} = \frac{c_n}{in}(|\zeta|^2 - it)^{-n}$$

in the sense of distributions because $\frac{c_n}{in}(|\zeta|^2 + \rho - it)^{-n}$ is dominated uniformly by the locally integrable function $\frac{c_n}{in}(|\zeta|^2 - it)^{-n}$. Taking derivatives,

$$\lim_{\rho \to 0} c_n(|\zeta|^2 + \rho - it)^{-n-1} = \frac{\partial}{\partial t}\left(\frac{c_n}{in}(|\zeta|^2 - it)^{-n}\right)$$

The right-hand side is a distribution of type $-a = -2n - 2$ that agrees with S_0 outside the origin. Therefore by Proposition 3′, S_0 has mean-value zero and $\lim_{\rho \to 0} S_\rho = c\delta + \text{p.v.}(S_0)$.

The constant c was computed by Koranyi and Vagi, [29,§6]. They showed that for fixed ρ,

(19) $$\lim_{N\to\infty} c_n \int_{||\zeta|^2 - it| \le N} S_\rho(\zeta, t)\, d\zeta\, dt = \frac{1}{2}$$

from which it follows that $c = \frac{1}{2}$.

3.3 Parametrices of $\Box_b$

Let $\zeta_j = x_j + iy_j$. The vector fields $X_j = \frac{\partial}{\partial x_j} + 2y_j \frac{\partial}{\partial t}$, $Y_j = \frac{\partial}{\partial y_j} - 2x_j \frac{\partial}{\partial t}$, and $T = \frac{\partial}{\partial t}$ form a basis for the left invariant vector fields on $\mathbb{H}^n$. They satisfy the commutation relations

$$[Y_j, X_j] = 4T \qquad j=1,\ldots,n,$$

and all other commutators vanish.

Denote $Z_j = \frac{1}{2}(X_j - iY_j)$ and $\overline{Z}_j = \frac{1}{2}(X_j + iY_j)$. The operators

$$\mathcal{L}_\alpha = -\frac{1}{2}\Sigma(Z_j\overline{Z}_j + \overline{Z}_j Z_j) + i\alpha T, \qquad \alpha \in \mathbb{C}$$

are analogous to the Laplacian on $\mathbb{R}^n$ in the sense that (up to constant multiples) they are the only operators that are homogeneous of degree -2 and invariant under $U(n)$ and left translation by $\mathbb{H}^n$.

One is also led to $\mathcal{L}_\alpha$ from complex analysis. Under the identification of $\mathbb{H}^n$ with $b\mathcal{D}$ given above, $\overline{Z}_j$ annihilates holomorphic functions restricted to $b\mathcal{D}$. We therefore introduce a boundary version of $\overline{\partial}$, called $\overline{\partial}_b$, defined on functions f by

$$\overline{\partial}_b f = \sum_{j=1}^{n} (\overline{Z}_j f) \cdot d\overline{\zeta}_j$$

$\overline{\partial}_b$ has a unique extension as a derivation mapping q-forms to q+1-forms. (We use the terminology "q forms" to mean forms which are expressions in the $d\bar{\zeta}_j$ of degree q.) We can define its formal adjoint $\bar{\partial}_b^*$: (q+1) forms $\to$ q-forms, using the fact that Z_j is the adjoint of $\bar{Z}_j$. (Here we have taken adjoints in $L^2(\mathbb{H}^n)$ with respect to Haar measure on the Heisenberg group.) The Kohn Laplacian is defined by

$$\Box_b = \bar{\partial}_b \bar{\partial}_b^* + \bar{\partial}_b^* \bar{\partial}_b$$

It takes q forms to q forms. If we denote by $\Box_b^{(q)}$ the restriction of $\Box_b$ to q-forms, then $\Box_b^{(q)} = \mathcal{L}_\alpha$, for $\alpha = n-2q$. (For this identity see Folland and Stein, [16, § 5].)

We now introduce distributions φ_α of class -2n = - a + 2 given by

$$\varphi_\alpha(\zeta, t) = (|\zeta|^2 - it)^{-\frac{(n+\alpha)}{2}} (|\zeta|^2 + it)^{-\frac{(n-\alpha)}{2}}$$

Notice that φ_α can be defined by analytic continuation for all $\alpha \in \mathbb{C}$ because $\operatorname{Re}(|\zeta|^2 \pm it) \geq 0$.

Theorem 6. $\mathcal{L}_\alpha \varphi_\alpha = c_\alpha \delta$, where $c_\alpha = 2^{2-n} \pi^{n+1} / \Gamma(\frac{n+\alpha}{2}) \Gamma(\frac{n-\alpha}{2})$. c_α is entire, but vanishes at the poles of the gamma function:

$$c_\alpha = 0 \text{ iff } \pm\alpha = n, n+2, n+4, \ldots$$

Thus $K_\alpha = c_\alpha^{-1} \varphi_\alpha$ is a fundamental solution for $\mathcal{L}_\alpha$ except for these special values of α.

It is not surprising that a fundamental solution fails to exist for certain values of α. For example, when $q = 0$, $\alpha = n$, we can see that $\Box_b^{(0)}$ is not hypoelliptic. In fact, $\Box_b^{(0)}$ annihilates boundary values of holomorphic functions, and it is easy to show that these need not be at all smooth. So $\square_b^{(0)}$ cannot have a fundamental solution that is smooth away from 0. On the other hand, when $0 < q < n$, $-n < \alpha < n$, so $\square_b^{(q)}$ does have a fundamental solution.

Proof sketch of Theorem 6

First let us point out that one is led to φ_α by first observing that the homogeneity and rotation invariance of $\mathcal{L}_\alpha$ imply that $\varphi_\alpha(\zeta, t)$ should have the form $\Phi(|\zeta|^2/t)t^{-n}$. One can then solve $\mathcal{L}_\alpha \varphi_\alpha = 0$ away from the origin more easily, as an ordinary differential equation.

To evaluate c_α, we consider the functions $\varphi_{\alpha,\rho}(\zeta, t) = (|\zeta|^2 + \rho - it)^{-(\frac{n+\alpha}{2})}(|\zeta|^2 + \rho + it)^{-(\frac{n-\alpha}{2})}$. $\lim_{\rho \to 0} \varphi_{\alpha,\rho} = \varphi_\alpha$ by dominated convergence.

One can compute that

$$\mathcal{L}_\alpha \varphi_{\alpha,\rho} = (\text{const})\,\rho\,(|\zeta|^2 + \rho - it)^{-(\frac{n+\alpha+2}{2})}(|\zeta|^2 + \rho + it)^{-(\frac{n-\alpha+2}{2})}.$$

The right-hand side clearly approaches zero if $(\zeta, t) \neq 0$ and $\rho \to 0$. Its integral turns out to be independent of ρ. So the evaluation of c_α reduces to the evaluation of

$$\int_{\mathbb{H}^n} \frac{d\zeta\, dt}{(|\zeta|^2 + 1 - it)^{\frac{n+\alpha+2}{2}}(|\zeta|^2 + 1 + it)^{\frac{n-\alpha+2}{2}}}.$$

It is possible that there is some connection between this computation and the previous one for S_ρ (see (19)).

There is a substitute identity in the case $\alpha = n$, $q = 0$. Recall

$$\mathcal{L}_n = \Box_b^{(0)} = -\sum_{j=1}^{n} Z_j \overline{Z}_j \,.$$

Since c_α has a simple zero at $\alpha = n$, it is natural to differentiate $\mathcal{L}_\alpha \varphi_\alpha = c_\alpha \delta$ with respect to α:

$$\mathcal{L}'_\alpha \varphi_\alpha + \mathcal{L}_\alpha \varphi'_\alpha = c'_\alpha \delta$$

Further calculation yields, (with $\alpha = n$)

$$\mathcal{L}_n K = \delta - C, \text{ where } C = \frac{1}{2}\delta + \text{p.v.}(S_0),$$

and $K = c_0\left[\log\left(\dfrac{|\zeta|^2 - it}{|\zeta|^2 + it}\right)\right](|\zeta|^2 - it)^{-n}$, $c_0 = \dfrac{2^{n-2}(n-1)!}{\pi^{n+1}}$. Notice that K is homogeneous of degree $-2n$ and that convolution with C is projection onto the null space of $\mathcal{L}_n$. Convolution with $\delta - C$ is projection onto the complement of the null space of $\mathcal{L}_n$.

Additional bibliographical comments

For Theorem 1, see Kree [31] and Gårding [17]. Theorem 2′ is taken from Knapp and Stein [28], but the proof given here is slightly simpler. For Theorem 3 and other derivations of Theorem 4, see Korányi and Vági [29].

Further details on the material in §3 are in Folland and Stein [16], and Greiner, Kohn, and Stein [19].

An excellent introduction and survey of analysis on nilpotent groups relevant to §2 and §3 may be found in the monograph of Goodman [18].

Chapter II. Basic Estimates for Pseudo Differential Operators

§4 Examples of symbols

We will now consider further the above examples in order to guess the appropriate hypotheses for a relevant class of pseudo differential operators. Recall that a pseudo differential operator denoted $a(x, D)$ is given by

$$f \longrightarrow \int_{\mathbb{R}^n} a(x,\xi)\, e^{-2\pi i x\xi}\, \hat{f}(\xi)\, d\xi$$

where $\hat{f}(\xi) = \int_{\mathbb{R}^n} e^{2\pi i x\xi} f(x)dx$ and $f \in \mathcal{S}$. We will assume that the symbol $a(x,\xi)$ is smooth everywhere. (It is harmless to throw away the singularity in ξ near the origin.) Since we are only concerned with local theorems, we will also assume that $a(x,\xi)$ has compact support in x.* Classical symbols $a(x,\xi) \in S^m_{1,0}$ satisfy

$$(1) \qquad |\partial_\xi^\alpha \partial_x^\beta a(x,\xi)| \le C_{\alpha,\beta}(1+\|\xi\|)^{m-|\alpha|},$$

where $\|\xi\|$ denotes Euclidean norm. More generally, $S^m_{\rho,\delta}$ is a similar class with $(1+\|\xi\|)^{m-|\alpha|}$ replaced by $(1+\|\xi\|)^{m-\rho|\alpha|+\delta|\beta|}$.

Up to constant factors, the heat kernel has symbol (see Example 2 in §1) $1/(\|\xi'\|^2 + i\xi_{n+1})$, where ξ_{n+1} is dual to the time variable and $\xi' = (\xi_1, \dots, \xi_n)$ is dual to the space variables. Directly related to this is the symbol $a(\xi) = \xi_j\xi_k/(\|\xi'\|^2 + i\xi_{n+1})$ where $j,k \le n$. $a(\xi)$ is homogeneous of degree zero under the dilations

$$\delta(\xi) = (\delta\xi_1, \dots, \delta\xi_n, \delta^2\xi_{n+1}).$$

*This assumption is made for convenience of formulation, but is not used essentially.

Let $|\xi| \approx \|\xi'\| + |\xi_{n+1}|^{\frac{1}{2}}$. Then

(2) $$\frac{\partial}{\partial \xi_j} a(\xi) \lesssim |\xi|^{-1} \lesssim \|\xi\|^{-\frac{1}{2}}, \quad j \leq n,$$

(3) $$\frac{\partial}{\partial \xi_{n+1}} a(\xi) \lesssim |\xi|^{-2} \lesssim \|\xi\|^{-1}.$$

The best we can say from an isotropic viewpoint is that $a \in S^0_{\frac{1}{2},0}$. The symbols of $S^0_{\frac{1}{2},0}$ act on L^2, but $b(\xi) = e^{i\|\xi\|^{\frac{1}{2}}}$ (ξ large) is a symbol in $S^0_{\frac{1}{2},0}$ that fails to act on L^p for $p \neq 2$. (See Hirschman [22] and Wainger [44].) Thus we need to exploit the stronger estimate (3). Fortunately, estimate (3) is actually <u>better</u> than the $S^0_{1,0}$ estimate. When $\|\xi'\|^2$ is much larger than ξ_{n+1}, $|\xi|^{-2}$ is strictly smaller than $\|\xi\|^{-1}$.

This observation is of capital importance for us. We shall develop this idea further by computing the symbols of the operators on the Heisenberg group discussed in §3 of the previous chapter.

In order to compute the symbol of a mapping $f \to \int_{\mathbb{H}^n} f(xy^{-1})k(y)dy = (f*k)(x)$, observe that for x and y in $\mathbb{H}^n$,

$$y^{-1}x = L_x(x-y)$$

where L_x is a linear transformation, $\det L_x = 1$ and $x \to L_x$ is smooth (even linear). In fact, the formula $(-\zeta', -t')(\zeta, t) = (\zeta - \zeta', t-t' + 2\,\mathrm{Im}((\zeta-\zeta')\bar{\zeta}))$ exhibits the operator L_x.

<u>Proposition 1.</u> $f \to f*k$ <u>is realized as a pseudo differential operator with symbol</u> $a(x,\xi) = \hat{k}(\tilde{L}_x(\xi))$, <u>where</u> $\tilde{L}_x = {}^t L_x^{-1}$.

Proof. Just change variables to $\eta = {}^tL_x\xi$;

$$f * k(x) = \int f(y)\,k(L_x(x-y))dy =$$

$$= \iint f(y)\, e^{-2\pi i L_x(x-y)\cdot\xi}\,\hat{k}(\xi)dy\,d\xi$$

$$= \iint f(y)\, e^{-2\pi i (x-y)\cdot\eta}\,\hat{k}(\tilde{L}_x\eta)dy\,d\xi$$

$$= \int \hat{f}(\eta)\, e^{-2\pi i x\cdot\eta}\,\hat{k}(\tilde{L}_x\eta)d\eta$$

Proposition 1 shows that we can find the symbol of the convolution operators of Section 3 of Chapter I by computing the Fourier transform of their kernels.

Recall that $S_\rho(\zeta,t) = c_n(|\zeta|^2 + \rho - it)^{-n-1}$. Let $\xi' = (\xi_1,\ldots,\xi_{2n})$ be dual to $\zeta = (x_1,\ldots,x_n,y_1,\ldots,y_n)$ and let ξ_{2n+1} be dual to t. It is easy to verify that the inverse Fourier transform of

$$\hat{S}_\rho(\xi) = \begin{cases} \text{const.}\ e^{-\pi\|\xi'\|^2/\xi_{2n+1}}\, e^{-2\pi\rho\xi_{2n+1}}, & \xi_{2n+1} > 0 \\ 0 & \text{otherwise} \end{cases}$$

is $S_\rho(\zeta,t)$. Therefore

$$\hat{S}_0(\xi) = \begin{cases} \text{const.}\ e^{-\pi\|\xi'\|^2/\xi_{2n+1}}, & \xi_{2n+1} > 0 \\ 0 & \text{otherwise} \end{cases}.$$

A more complicated example (occurring for $\Box_b$) is

$$K_\alpha(\zeta,t) = c_\alpha^{-1}(|\zeta|^2 - it)^{-\left(\frac{n+\alpha}{2}\right)}(|\zeta|^2 + it)^{-\left(\frac{n-\alpha}{2}\right)}.$$

$$\hat{K}_\alpha(\xi) = \text{const.}\ \frac{1}{\xi_{2n+1}} \int_0^1 (1-s)^{\frac{n-\alpha}{2}-1} (1+s)^{\frac{n+\alpha}{2}-1} e^{-s(\|\xi'\|^2/\xi_{2n+1})} ds,$$

for $\xi_{2n+1} > 0$. (For $\xi_{2n+1} < 0$, we must interchange α and $-\alpha$; see Greiner-Stein [20].) Notice that K_α is homogeneous of degree $-2n$ and $\hat{K}_\alpha$ is homogeneous of degree $-2 = -2n-2-(-2n)$.

As with the heat equation we can see from the expression $e^{-\|\xi'\|^2/\xi_{2n+1}}$ that appears above, that $|\xi| \simeq \|\xi'\| + |\xi_{2n+1}|^{\frac{1}{2}}$ is the right kind of norm to use. However, in light of Proposition 1, the precise norm should vary with x in the following way

$$\rho(x, \xi) = |\tilde{L}_x(\xi)|.$$

We can verify, as in the case of the heat equation, that a derivative in a "good" (= non-characteristic) direction gives a gain like $1/\rho(x,\xi)$ and a derivative in a "bad" (= characteristic) direction gives a gain like $1/\rho(x,\xi)^2$. Also, when ξ tends to infinity in a good direction $\rho(x,\xi) \approx \|\xi\|$, but when ξ tends to infinity in a bad direction $\rho(x,\xi) \approx \|\xi\|^{\frac{1}{2}}$.

A definition that incorporates these various properties in a single inequality is as follows. Let η be a vector in $\mathbb{R}^n$, and denote by $(\eta, \frac{\partial}{\partial \xi})$ the operator $\sum \eta_j \frac{\partial}{\partial \xi_j}$. The symbols $a(x,\xi)$ in the class S^m_ρ will satisfy estimates like

$$|a(x,\xi)| \leq C\rho(x,\xi)^m \quad \text{for large } \xi$$

$$(4) \qquad |(\eta, \frac{\partial}{\partial \xi})\, a(x,\xi)| \leq C\rho(x,\xi)^m \left(\frac{\rho(x,\eta)}{\rho(x,\xi)} + \frac{\rho(x,\eta)^2}{\rho(x,\xi)^2}\right)$$

$$\text{for large } \xi,\ \|\eta\| \geq 1.$$

If η is a unit vector in any direction, then the right-hand side of (4) is bounded by $C\rho^{m-1}$. But if η points in a bad direction, $\rho(x,\eta) \approx \|\eta\|^{\frac{1}{2}}$; so by letting η tend to infinity, we can obtain a stronger estimate $C\rho^{m-2}$. In directions that are a mixture of good and bad we get some kind of intermediate estimate.

This is only a preview of the kinds of conditions we will impose on our symbols. To understand derivatives in x, we look at expressions like $\frac{\partial}{\partial x_j}\hat{k}(\tilde{L}_x(\xi))$ and $\frac{\partial}{\partial \xi_k}\frac{\partial}{\partial x_j}\hat{k}(\tilde{L}_x(\xi))$. These are more complicated and will be postponed until later.

§5 The distance function $\rho(x,\xi)$

We now start anew with the general case. We first introduce a distance function $\rho(x,\xi)$ for the (cotangent) ξ-space at each point x. We then define a dual notion of distance on the (tangent) λ-space, which can be identified with a distance in the x-space itself. Our task is to show that this (second) distance satisfies the properties for the Vitali covering lemma.

Let $Q_x^1(\xi)$ be a positive semi-definite quadratic form in ξ depending smoothly on x. We will allow the signature of Q_x^1 to vary as x varies. In applications to operators of the form $\sum a_{ij}(x)\frac{\partial^2}{\partial x_i \partial x_j} + \ldots$, Q_x^1 will be the form with matrix $\{a_{ij}(x)\}$.

Example 1. In $\mathbb{R}^2$, $\frac{\partial^2}{\partial x^2} + x^2\frac{\partial^2}{\partial t^2}$. This Grushin-type operator degenerates at $x=0$. The form is $\xi^2 + x^2\tau^2$, with (ξ,τ) dual to (x,t).

Example 2. (Kannai) $t\frac{\partial^2}{\partial x^2} \pm \frac{\partial}{\partial t}$. For $t > 0$, the form $t\xi^2$ is positive semi-definite, but for $t < 0$ it is negative semi-definite.

To take care of Example 2 we need to define $Q(x, \xi) = \varphi(x)\, Q^1_x(\xi)$, where $\varphi(x)$ is a smooth real-valued function with the properties

$$\varphi(x) = 0 \Rightarrow d\varphi(x) \neq 0 \tag{5}$$

$$\varphi(x) = 0 \Rightarrow Q^1_x(d\varphi(x)) = 0 . \tag{6}$$

Later on we will add the hypothesis that $Q^1_x(\xi)$ can be written as a sum of squares smoothly in x, i.e.,

$$Q^1_x(\xi) = \sum_{j=1}^{N} |L^j_x(\xi)|^2, \quad (L^j_x \text{ linear}).$$

This is stronger than the assumption that Q^1_x is positive semi-definite.

Definition. $\rho(x, \xi) = (Q(x, \xi)^2 + \|\xi\|^2)^{1/4}$, when $\|\xi\| \geq 1$.

The dual distance is defined for $\lambda \in \mathbb{R}^n$ by

Definition. $N_x(\lambda) = \sup \{\frac{1}{\rho(x, \xi)} : |(\lambda, \xi)| \geq 1\}$. Here (λ, ξ) is the ordinary inner product.

We will only be interested in ξ large and λ small. Denote

$$B^*(x, R) = \{\xi: \rho(x, \xi) < R\}, \quad R \geq 1$$

$$B(x, \delta) = \{\lambda: N_x(\lambda) < \delta\}, \quad \delta \leq 1.$$

Finally, we need one more notion to keep track of the relation between Euclidean distance and our distance. Denote

$$V^*_x(R) = \text{Euclidean volume of } B^*(x, R)$$

$$V_x(\delta) = \text{Euclidean volume of } B(x, \delta).$$

We will say that $A(x,\xi) \lesssim B(x,\xi)$ if there is a constant C independent of (x,ξ) for x in a compact set such that

$$A(x,\xi) \le C\, B(x,\xi)$$

(We will use similar notations for functions of (x,λ), and give the obvious meaning to $\simeq$.)

Here are four simple propositions.

<u>Proposition 2.</u> (a) $\|\xi\|^{\frac{1}{2}} \lesssim \rho(x,\xi) \lesssim \|\xi\|$ $\quad \xi$ <u>large</u>

(b) $\|\lambda\| \lesssim N_x(\lambda) \lesssim \|\lambda\|^{\frac{1}{2}}$ $\quad \lambda$ <u>small</u>.

<u>Remark.</u> $\rho(x, r\xi)$ <u>is an increasing function of</u> r.

<u>Proposition 3.</u> (a) $\rho(x,\xi_1+\xi_2) \lesssim \rho(x,\xi_1) + \rho(x,\xi_2)$

(b) $N_x(\lambda_1+\lambda_2) \lesssim N_x(\lambda_1) + N_x(\lambda_2)$

<u>Proposition 4.</u> $V_x^*(R)^{-1} \simeq V_x(\delta)$ where $\delta = 1/R$

<u>Proposition 5.</u> (a) $V_x(2\delta) \lesssim V_x(\delta)$

(b) $V_x(t\delta) \lesssim t^{2n} V(\delta)$, $\quad t \ge 1$

(c) $V_x(t\delta) \lesssim t^{n/2} V(\delta)$, $\quad t \le 1$.

Proposition 2(a) is trivial, and 2(b) follows by duality. Proposition 3 follows from the fact that Schwarz' inequality is valid for semi-definite forms. Proposition 4 is not much harder. It may be simplest to observe that the ball $B^*(x,R)$ is comparable to the "ellipse"

$$\widetilde{B}^*(x,R) = \{\xi; R^{-2}|Q(x,\xi)| + R^{-4}\|\xi\|^2 \le 1\}$$

based on the (strictly) positive definite form $R^{-2}|Q(x,\xi)| + R^{-4}\|\xi\|^2$. Thus Proposition 4 follows from the more familiar notion of balls and dual balls for positive definite forms. Proposition 5b) and c) follow from Proposition 4 and the inclusions $B^*(tR) \subset t^2 B^*(R)$, $t \geq 1$, and $B^*(tR) \subset t^{\frac{1}{2}} B^*(R)$, $t \leq 1$. Proposition 5a) is just a special case of 5b).

Examples of $B(x,\delta)$

1. $\rho \simeq \|\xi\|$

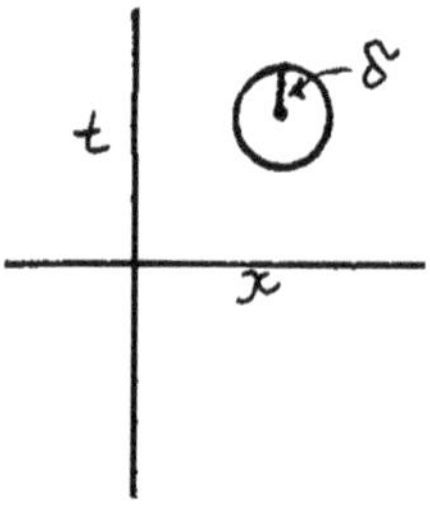

2. Heat equation, $Q(\xi,\tau) = \xi^2$

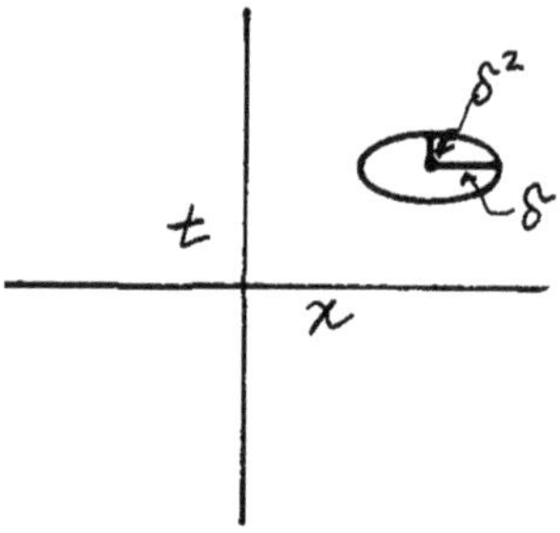

3. $\dfrac{\partial^2}{\partial x^2} + x^2 \dfrac{\partial^2}{\partial t^2}$

$Q((x,t),(\xi,\tau)) = \xi^2 + x^2\tau^2$

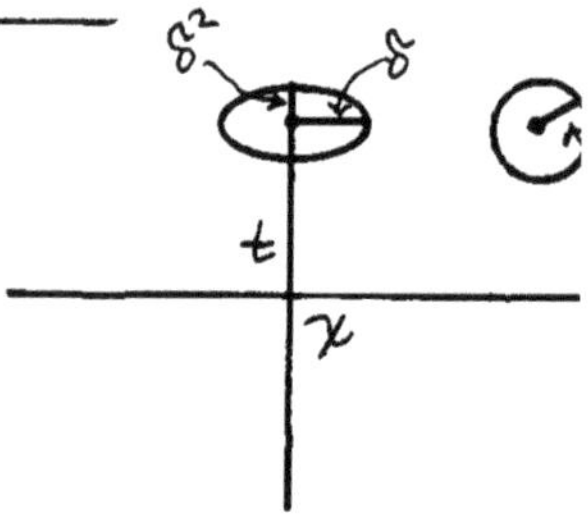

4. $t\dfrac{\partial^2}{\partial x^2} \pm \dfrac{\partial}{\partial t}$

$Q((x,t),(\xi,\tau)) = t\xi^2$

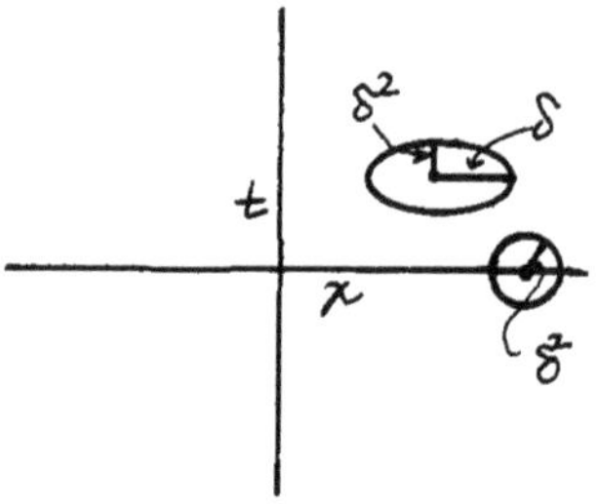

So far we have not used very special properties of $\rho(x,\xi)$ For example,

$$\rho(x,\xi) = \sum_{j=1}^{N-1} |Q_x^j(\xi)|^{a_j} + \|\xi\|^{a_N}$$ also satisfies similar properties.

Our next goal is to prove that N_x defines a distance in the x space

(Proposition 9) and that the distance satisfies an approximate triangle law (Proposition 12). In the process we will use the special form of $\rho(x,\xi)$ in a more essential way.

Proposition 6. $|(\lambda,\xi)| \lesssim N_x(\lambda)|Q(x,\xi)|^{\frac{1}{2}} + N_x(\lambda)^2\|\xi\|.$

Proof. Denote $t = |(\lambda,\xi)|$ and $\xi_0 = t^{-1}\xi$. Then $|(\lambda,\xi_0)| = 1$, and by definition

$$N_x(\lambda) \geq \frac{1}{\rho(x,\xi_0)} .$$

In other words,

$$1 \leqq N_x(\lambda)\rho(x,\xi_0) \lesssim N_x(\lambda)|Q(x,\xi_0)|^{\frac{1}{2}} + N_x(\lambda)\|\xi_0\|^{\frac{1}{2}}.$$

Thus

$$t \lesssim N_x(\lambda)|Q(x,\xi)|^{\frac{1}{2}} + N_x(\lambda)\|\xi\|^{\frac{1}{2}}t^{\frac{1}{2}}.$$

Let $a = N_x(\lambda)\|\xi\|^{\frac{1}{2}}$, $b = t^{\frac{1}{2}}$, and use $ab \leq \frac{1}{2}a^2 + \frac{1}{2}b^2$ to conclude the proof.

Proposition 7. $|(\nabla_x Q(x,\xi),\lambda)| \lesssim \|\xi\|\rho(x,\xi)N_x(\lambda) + \|\xi\|^2 N_x(\lambda)^2.$

Lemma 1. $|\nabla_x Q_x^1(\xi)| \lesssim Q_x^1(\xi)^{\frac{1}{2}}\|\xi\|.$

Lemma 1 will be proved using

Lemma 2. Suppose $f \in C^2[-2,2]$, $f \geq 0$. Then $|f'(t)| \leq C|f(t)|^{\frac{1}{2}}$ for $t \in [-1,1]$, where C depends only on the C^2 norm of f.

Proof. Let $t \in [-1,1]$; then

(7) $$f(t+h) - f(t) = hf'(t) + O(h^2).$$

Case (i). $|f(t)| \geq 1$. This case is trivial.

Case (ii). $|f(t)| < 1$ and $f'(t) \leq 0$.

Let $h = |f(t)|^{\frac{1}{2}}$. Then (7) implies

$$hf'(t) \geq -f(t) - O(h^2), \quad \text{since } f(t+h) \geq 0.$$

Therefore $\quad |f'(t)| \leq C|f(t)|^{\frac{1}{2}}$ as desired.

Case (iii). $|f(t)| < 1$ and $f'(t) > 0$. Treat this case in the same way as case (ii) with $h = -|f(t)|^{\frac{1}{2}}$.

To prove Lemma 1, notice that we may assume without loss of generality that $\|\xi\| = 1$. The lemma then follows from Lemma 2.

Proof of Proposition 7. Let F denote the zero set of φ. For $x \in F$, $d\varphi(x) \neq 0$. So on any compact subset we have the estimate $|\varphi(x)| \gtrsim \text{dist}(x,F)$. Since $Q_x^1(d\varphi(x)) = 0$ for $x \in F$, it follows that

$$|Q_x^1(d\varphi(x))| \lesssim |\varphi(x)|, \tag{8}$$

for x near F. But (8) is also valid trivially for x far from F, so it holds in general.

We compute

$$\nabla_x Q(x,\xi) = (\nabla\varphi(x))\, Q_x^1(\xi) + \varphi(x)\, \nabla_x Q_x^1(\xi). \tag{9}$$

By Lemma 1,

$$\begin{aligned} |(\varphi(x)\nabla_x Q_x^1(\xi), \lambda)| &\lesssim \|\lambda\| \; |\varphi(x)| \; |Q_x^1(\xi)|^{\frac{1}{2}} \; \|\xi\| \\ &\lesssim \|\lambda\| \; |\varphi(x)|^{\frac{1}{2}} \; |Q_x^1(\xi)|^{\frac{1}{2}} \; \|\xi\| \\ &\lesssim \|\lambda\| \; \|\xi\| \; |Q(x,\xi)|^{\frac{1}{2}} \\ &\lesssim N_x(\lambda) \; \|\xi\| \; \rho(x,\xi) \end{aligned}$$

The last step follows from Proposition 2: $\|\lambda\| \lesssim N_x(\lambda)$.

For the first term of (9):

$$|\nabla\varphi(x)\, Q_x^1(\xi), \lambda)| = |Q_x^1(\xi)|\ |(\nabla\varphi(x), \lambda)|$$

$$\lesssim |Q_x^1(\xi)|\ (N_x(\lambda)\,|Q(x, \nabla\varphi(x))|^{\frac{1}{2}} + N_x(\lambda)^2)$$

by Proposition 6 (with $\xi = \nabla\varphi(x)$). Now apply (8):

$$\lesssim |\varphi(x)\, Q_x^1(\xi)|\, N_x(\lambda) + Q_x^1(\xi)\, N_x(\lambda)^2$$

$$\lesssim \rho(x, \xi)^2\, N_x(\lambda) + \|\xi\|^2\, N_x(\lambda)^2$$

$$\lesssim \|\xi\|\, \rho(x, \xi)\, N_x(\lambda) + \|\xi\|^2\, N_x(\lambda)^2.$$

<u>Proposition 8.</u> $|(\nabla_x \rho(x, \xi), \lambda)| \lesssim \|\xi\|\, N_x(\lambda) + \dfrac{\|\xi\|^2}{\rho(x, \xi)}\, N_x(\lambda)^2.$

<u>Proof.</u> $\rho^4 = Q^2 + \|\xi\|^4$ implies that $(4\rho^3\, \nabla_x \rho, \lambda) = (2\, Q\, \nabla_x\, Q, \lambda)$. Now use $Q \leq \rho^2$ and Proposition 7.

<u>Proposition 9.</u> $N_x(x-y) \simeq N_y(x-y)$.

<u>Proof.</u> $N_x(x-y) = \sup\limits_{\xi} \{ \dfrac{1}{\rho(x, \xi)} : |(x-y, \xi)| = 1 \}.$

Fix a vector ξ where the supremum is attained. $N_x(y-x) = \rho(x, \xi)^{-1}$. By Taylor's formula, and Propositions 2 and 7,

$$Q(y, \xi) \lesssim Q(x, \xi) + (\nabla_x Q(x, \xi), y-x) + O(\|\xi\|^2\, \|y-x\|^2)$$

$$\lesssim \rho(x, \xi)^2 + \|\xi\|\, \rho(x, \xi)\, N_x(y-x) + \|\xi\|^2\, N_x(y-x)^2$$

$$\lesssim \rho(x, \xi)^2$$

Therefore, $\rho(y, \xi) \lesssim \rho(x, \xi)$. Finally,

$$N_y(x-y) = \sup \{ \frac{1}{\rho(y, \xi)} : |(x-y, \xi)| = 1 \} \gtrsim N_x(x-y).$$

The opposite inequality holds by symmetry.

Corollary. Let $x_t = tx + (1-t)y$, $0 \leq t \leq 1$. Then $N_{x_t}(x-y) \simeq N_x(x-y)$.

Proof. Without loss of generality $t \leq 1/2$. There is a number r, $1 < r < 2$ such that

$$N_{x_t}(x-y) = N_{x_t}(r(x-x_t)) \simeq N_{x_t}(x-x_t) \simeq N_x(x-x_t) \simeq N_x(x-y)$$

(The outer equivalences use a simple consequence of Proposition 3b): $N_x(2\lambda) \approx N_x(\lambda)$.)

Proposition 10. $\dfrac{\rho(y,\xi)}{\rho(x,\xi)} \lesssim 1 + N_x(x-y)^{2/3} \rho(y,\xi)^{2/3}$.

Proof. First assume that $\rho(y,\xi) \geq \rho(x_t,\xi)$ for all $t \in [0,1]$. Using successively the mean-value theorem, Proposition 8, and the corollary to Proposition 9, we obtain

$$\left|\frac{\rho(y,\xi)^3 - \rho(x,\xi)^3}{\rho(x,\xi)^3}\right| \leq$$

$$\frac{1}{\rho(x,\xi)^3} 3\rho(x_t,\xi)^2 \, |(\nabla_x \rho(x_t,\xi), x-y)|$$

$$\lesssim \frac{\rho(x_t,\xi)^2}{\rho(x,\xi)^3}\left(\|\xi\| N_{x_t}(x-y) + \frac{\|\xi\|^2}{\rho(x_t,\xi)} N_{x_t}(x-y)^2\right)$$

$$\lesssim \frac{\rho(y,\xi)^2}{\rho(x,\xi)} N_x(x-y) + \rho(y,\xi)^2 N_x(x-y)^2.$$

Suppose that $\frac{1}{p} + \frac{1}{q} = 1$. It is easy to show that

(10) $AB \leqq \epsilon A^p + C(\epsilon) B^q$, for any $\epsilon > 0$.

For $p = 3$ and $q = 3/2$, $A = \frac{\rho(y,\xi)}{\rho(x,\xi)}$ and $B = \rho(y,\xi)\, N_x(x-y)$ we can apply (10) to the above:

$$\frac{\rho(y,\xi)^3}{\rho(x,\xi)^3} \lesssim 1 + \rho(y,\xi)^{3/2} N_x(x-y)^{3/2} + \rho(y,\xi)^2 N_x(x-y)^2$$

$$\lesssim 1 + \rho(y,\xi)^2 N_x(x-y)^2 .$$

To remove the assumption that $\rho(y,\xi) \geq \rho(x_t,\xi)$, pick the point y_0 closest to x such that $\rho(y_0,\xi) = \rho(y,\xi)$. By the intermediate value theorem $\rho(y_0,\xi) \geq \rho(x_t,\xi)$ for all x_t between y_0 and x, and we are reduced to the previous case.

Remark. *In the case $\varphi(x) \equiv 1$, we could have obtained an improved* upper bound in Proposition 10, namely, $1 + N_x(x-y)^{1/2} \rho(y,\xi)^{1/2}$. This is typical of the extra work and weaker estimates that hold when φ is not necessarily constant.

Proposition 11. $N_x(\lambda) \lesssim N_y(\lambda) + N_x(x-y)^{2/3} N_y(\lambda)^{1/3}$.

[Compare this with Rothschild-Stein [39], Proposition 12.3]

Proof. Choose ξ such that $N_x(\lambda) = \frac{1}{\rho(x,\xi)}$ and $|(\xi,\lambda)| = 1$.

By Proposition 10,

$$N_x(\lambda) = \frac{1}{\rho(x,\xi)} \lesssim \rho(y,\xi)^{-1} + N_x(x-y)^{2/3} \rho(y,\xi)^{-1/3}$$

$$\lesssim N_y(\lambda) + N_x(x-y)^{2/3} N_y(\lambda)^{1/3}.$$

Proposition 12. (Triangle inequality)

$$N_x(x-y) \lesssim N_x(x-z) + N_z(z-y) .$$

Proof. Recall from Proposition 3 that

$$N_x(x-y) \lesssim N_x(x-z) + N_x(z-y)$$

Also, $N_x(z-y) \lesssim N_z(z-y) + N_x(x-z)^{2/3} N_z(z-y)^{1/3}$.

Now use (10):

$$N_x(x-z)^{2/3} N_z(z-y)^{1/3} \leq \varepsilon N_x(x-z) + C(\varepsilon) N_z(z-y)$$

to conclude the proof.

Proposition 13. There exists $M > 0$ such that

$$V_y(\delta) \lesssim V_x(\delta)\left(1 + \left(\frac{N_x(x-y)}{\delta}\right)^M\right)$$

Proof. Recall Proposition 5b)

$$V_x(t\delta) \lesssim t^{2n} V_x(\delta) \qquad t \geq 1.$$

By Proposition 12, $B(y,\delta) \subset B(x, C(\delta + N_x(x-y)))$. Therefore,

$$V_y(\delta) \leq V_x(C(\delta + N_x(x-y))) \lesssim V_x(\delta)\left(1 + \left(\frac{N_x(x-y)}{\delta}\right)\right)^{2n}.$$

We now introduce an abuse of our earlier notation, by setting

$$B(x,\delta) = \{y: N_x(x-y) + N_y(x-y) < \delta\}.$$

Thus a (symmetrized) ball about the origin in the tangent space at x is identified with a ball centered at x in the x-space.

These balls satisfy the properties

(11) $B(x,\delta)$ is open and bounded

(12) $B(x,\delta)$ is monotonic increasing in δ

(13) $m(B(x, 2\delta)) \leq C\, m(B(x, \delta))$

(14) There is a constant C such that if $B(x_1, \delta_1) \cap B(x_2, \delta_2) \neq \emptyset$ and $\delta_1 \geq \delta_2$, then $B(x_1, C\delta_1) \supseteq B(x_2, \delta_2)$.

Property (14) is a simple consequence of the triangle inequality. The assertion (13) follows from Proposition 5. The rest are obvious.

Lemma. A family of balls satisfying (11) - (14) also satisfies a Vitali covering property: If $B_1, \ldots, B_N$ are balls from this family, then there is a disjoint subcollection $B_{i_1}, \ldots, B_{i_k}$ such that $m(\bigcup_{j=1}^{N} B_j) \leq C \sum_{j=1}^{k} m(B_{i_j})$.

The lemma is proved using (11) to (14) by standard arguments. We refer the reader to Coifman and Weiss [10] for a general formulation in the setup of "quasi-homogeneous spaces."

One might hope to prove the Vitali property for balls arising from more general functions $\rho(x, \xi)$. But this fails even for a simple variant like $\rho(x, \xi) \approx |Q_x^1(x, \xi)|^{1/2} + \|\xi\|^{1/3}$, unless we place extra conditions on the derivatives in x of Q^1 in special directions. The conditions are far from coordinate free and seem to require "normal" coordinates for an intrinsic definition.

To illustrate this let us repeat the proof of (14) in a simple case. Suppose $B(x, \delta)$ is an ellipse in $\mathbb{R}^2$ about x with major axis of length δ, minor axis of length δ^2. Suppose that the direction of the axes varies smoothly with x. If $B(x_1, \delta) \cap B(x_2, \delta) \neq \emptyset$, then $|x_1 - x_2| < 2\delta$. Therefore, the major axes of $B(x_1, \delta)$ and $B(x_2, \delta)$ make an angle of at most $C\delta$. Let ℓ_1 be the line containing the major axis of $B(x_1, \delta)$. Since $B(x_2, \delta)$

has length δ, dist $(x, \ell_1) \leq C\delta^2$ for all $x \in B(x_2, \delta)$. Thus $B(x_2, \delta)$ is contained in a constant multiple of $B(x_1, \delta)$.

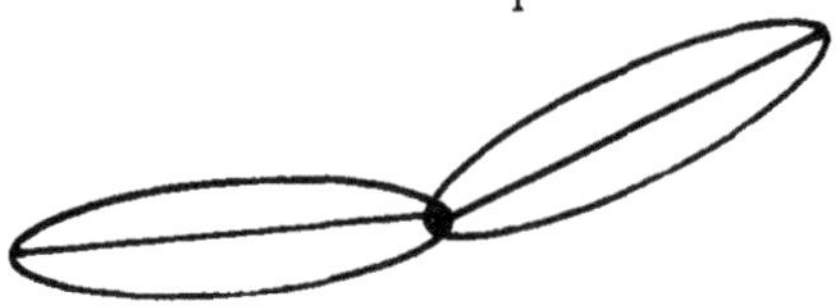

It is clear that (14) fails for ellipses with major and minor axes of length δ and δ^3, say, unless we impose additional conditions on how fast the orientation can vary when the center moves in the direction of the major axis. Moreover, unlike the previous case, a smooth change of variables preserving the center of such an ellipse does not even preserve its general shape, (i.e., up to bounded dilations).

What we have described here are some of the main obstacles standing in the way of an immediate extension of the theory, as presented below, to the "higher step" case. The ideas needed to surmount these obstacles are now better understood and are described in Nagel and Stein [35].

§6. L^p estimates ($p \neq 2$)

The Vitali covering lemma implies

Theorem 7. *Denote* $Mf(x) = \sup_{\delta > 0} \frac{1}{m(B(x,\delta))} \int_{B(x,\delta)} |f(y)|dy$. $f \rightarrow Mf$ *is bounded on* L^p $1 < p \leq \infty$ *and weak-type* (1,1).

The L^p theory for our pseudo-differential operators is obtained by adapting the methods of Calderón-Zygmund, using Theorem 7 as our starting point. For the general formulations relevant here see Coifman-Weiss [10], Korányi-Vagi [29], and Rivière [38]. The $\mathbb{R}^n$ theory is in Stein [41].

Lemma (Korányi-Vagi version) Given $f \geq 0$, $f \in L^1$ and $\alpha > 0$, there exists balls $B_n = B(x_n, \delta_n)$, $B_n^* = B(x_n, c\delta_n)$, and sets Q_n such that $B_n \subset Q_n \subset B_n^*$ and

(a) The sets Q_n are disjoint

(b) $c_1 \alpha \leq \frac{1}{m(Q_n)} \int_{Q_n} f(x) \leq c_2 \alpha$

(c) $f(x) \leqq \alpha$ for almost every $x \in {}^c\cup Q_n$

(d) $m(\cup Q_n) \leqq \frac{c}{\alpha} \|f\|_{L^1}$. (This follows from a) and b).)

Proof sketch. For each x such that $Mf(x) > \alpha$ consider the ball $B(x, r)$ for the smallest r such that

$$\frac{1}{m(B(x,cr))} \int_{B(x,cr)} f(y)dy \leqq \alpha$$

Select a disjoint subcollection $\{B_n\}$ of these balls by the (infinite version of) Vitali lemma. It is not hard to choose Q_n so that $B_n \subset Q_n \subset B_n^*$, the Q_n are disjoint, and $\cup Q_n \supset \{x: Mf(x) > \alpha\}$.

One can deduce from this lemma

Theorem 8. Suppose T is bounded on L^1 and

(i) $\|Tf\|_{L^2} \leqq A\|f\|_{L^2}$ for all $f \in L^1 \cap L^2$

(ii) $\int_{{}^cB(x, c\delta)} |Tf(y)|dz \leqq A \int_{B(x,\delta)} |f(y)|dz$ for all functions f supported on

$B(x,\delta)$ with $\int_{B(x,\delta)} f(y)dy = 0$. Then T is weak-type $(1,1)$ and bounded on L^p, $1 < p \leq 2$, with bounds depending only on A and p.

Let $Tf(x) = \int k(x,y)f(y)dy$. A sufficient condition for Theorem 8 to apply is

(i) $$\|Tf\|_{L^2} \leq A\|f\|_{L^2}$$

(ii)′ $$\int_{x\in {}^c B(y_0, c\delta)} |k(x,y) - k(x,y_0)|dx \leq A \quad \text{for } y \in B(y_0,\delta).$$

We now introduce a preliminary class of symbols.

Definition. The class $\tilde{S}^m_\rho$ is the collection of symbols $a(x,\xi) \in C^\infty(\mathbb{R}^n \times \mathbb{R}^n)$ with compact support in x satisfying for all $\|\eta_j\| \geq 1$, $\|\lambda_i\| \leq 1$.

(15) $$\left|(\eta_1, \frac{\partial}{\partial\xi}) \cdots (\eta_k, \frac{\partial}{\partial\xi})(\lambda_1, \frac{\partial}{\partial x}) \cdots (\lambda_\ell, \frac{\partial}{\partial x}) a(x,\xi)\right| \leq$$

$$\leq C_{k,\ell}\,\rho(x,\xi)^m \prod_{j=1}^{k} \theta\left(\frac{\rho(x,\eta_j)}{\rho(x,\xi)}\right) \prod_{i=1}^{\ell} \theta(\rho(x,\xi)\,N_x(\lambda_i))$$

where $\theta(t) = t + t^2$.

Theorem 9. Let $a \in \tilde{S}^0_\rho$ and

$$Tf(x) = \int a(x,\xi)\,e^{-2\pi i x \xi}\,\hat{f}(\xi)d\xi, \quad f \in \mathcal{S}$$

Suppose that T is bounded on L^2, then T is bounded on L^p, $1 < p < \infty$.

Remark. Theorem 9 remains true and the proof is unchanged if the function θ appearing twice in (15) is replaced by a pair of functions θ_1 and θ_2 each satisfying $\theta_j(t) \leq t^\epsilon$, for small t and some fixed ϵ, $\theta_j(t) \leq t^N$, for large t and some fixed N.* However, $\theta(t) = t + t^2$ is the

*In this connection compare Beals [2].

function that arises in all our applications.

In order to prove Theorem 9 we will make estimates on the kernel of T. Let us work formally for the moment. Denote $K(x,z) = \int a(x,\xi)e^{-2\pi i\xi\cdot z}d\xi$. Then

$$a(x,D)\,f(x) = \int K(x,x-y)\,f(y)dy$$

Lemma 1. For $\|z\| < 1$

(a) $|K(x,z)| \lesssim V_x(\delta)^{-1}$ where $\delta = N_x(z)$

(b) $|(\lambda,\frac{\partial}{\partial z})K(x,z)| \lesssim V_x(\delta)^{-1}\theta(N_x(\lambda)/N_x(z))$, $\|\lambda\| \leq 1$

(c) $|(\lambda,\frac{\partial}{\partial x})K(x,z)| \lesssim V_x(\delta)^{-1}\theta(N_x(\lambda)/N_x(z))$, $\|\lambda\| \leq 1$

To give a more rigorous meaning to the lemma, and because of further applications, we decompose $a(x,\xi)$. Choose a function $\psi \in C_0^\infty$ such that $\psi \geq 0$,

$$\psi(t) = \begin{cases} 1 & t \leq 1 \\ 0 & t \geq 2 \end{cases}$$

Denote $a_0(x,\xi) = \psi(\rho(x,\xi))\,a(x,\xi)$,

$$a_j(x,\xi) = [\psi(2^{-j}\rho(x,\xi)) - \psi(2^{-j+1}\rho(x,\xi))]\,a(x,\xi).$$

It follows that

$$a(x,\xi) = \sum_{j=0}^{\infty} a_j(x,\xi)$$

and for $j \geq 1$, $a_j(x,\xi)$ is supported in an annulus in ξ where $\rho(x,\xi) \simeq 2^j$. Define $K_j(x,z) = \int a_j(x,\xi)\,e^{-2\pi i\xi\cdot z}d\xi$. This definition is no longer merely formal.

Lemma 2. *If* $\|z\| \leq 1$

(a) $|K_j(x,z)| \leq C_M V_x(2^{-j})^{-1}(2^j N_x(z))^{-M}$ *for any* $M \geq 0$

(b) $|(\lambda, \frac{\partial}{\partial z}) K_j(x,z)| \leq C_M (V_x(2^{-j}))^{-1}(2^j N_x(z))^{-M}\theta(2^j N_x(\lambda))$

for $M \geq 0$, $\|\lambda\| \leq 1$

(c) $|(\lambda, \frac{\partial}{\partial x}) K_j(x,z)|$ *has an estimate identical with* b).

Definition. We will denote by $\hat{S}^m_\rho$ the class of symbols satisfying the $\tilde{S}^m_\rho$ conditions for derivatives in ξ only. A key point in estimating the kernel K_j is

Lemma 3.

(a) $\rho(x,\xi)^z \in \hat{S}^m_\rho$, *with* $m = \mathrm{Re}\, z$.

(b) *Under the additional hypothesis that* $Q^1_x(\xi)$ *is the sum of squares of linear forms*,

$$\rho(x,\xi)^z \in \tilde{S}^m_\rho, \quad \textit{where } m = \mathrm{Re}\, z.$$

Lemma 3b) shows in particular that the symbol class $\tilde{S}^m_\rho$ is not empty. From now on we will assume that $Q^1_x(\xi)$ is a sum of squares of linear forms.

For simplicity, we will only sketch the main point in the proof of Lemma 3b) when $\varphi(x) = 1$, namely that $Q_x(\xi) \in \tilde{S}^2_\rho$.* The most difficult part is to take one derivative in ξ and one in x. In fact,

$$Q_x(\xi) = \Sigma_j L^j_x(\xi)^2. \qquad \text{Therefore}$$

$$(\eta, \frac{\partial}{\partial \xi}) Q_x(\xi) = 2\Sigma_j L^j_x(\eta) L^j_x(\xi).$$

*The complete proof is presented in the Appendix.

Hence

$$|(\lambda, \frac{\partial}{\partial x})(\eta, \frac{\partial}{\partial \xi})Q_x(\xi)| = |2\Sigma(L_x^j(\eta)\cdot \lambda)\, L_x^j(\xi) + L_x^j(\eta)(L_x^j(\xi)\cdot\lambda)|$$

$$\lesssim (\|\eta\| \ \|\lambda\| \rho(x,\xi) + \|\xi\| \ \|\lambda\| \rho(x,\eta)).$$

We must dominate this with

$$\rho(x,\xi)^2 \, \theta\left(\frac{\rho(x,\eta)}{\rho(x,\xi)}\right) \theta\,(\rho(x,\xi)\, N_x(\lambda)).$$

In fact we shall only need

$$\rho(x,\xi)^2 \left(\frac{\rho(x,\eta)}{\rho(x,\xi)} + \frac{\rho(x,\eta)^2}{\rho(x,\xi)^2}\right) \rho(x,\xi)\, N_x(\lambda).$$

(The term with the factor $\rho(x,\xi)^2 N_x(\lambda)^2$ is not needed because we have considered $Q_x(\xi)$ instead of $\varphi(x)\, Q_x^1(\xi)$.) The proof is now completed with the help of

$$\|\eta\| \ \|\lambda\| \ \rho(x,\xi) \lesssim \rho(x,\eta)^2 \ N_x(\lambda)\, \rho(x,\xi) \quad \text{and}$$

$$\|\xi\| \ \|\lambda\| \ \rho(x,\eta) \lesssim \rho(x,\xi)^2 \ N_x(\lambda)\, \rho(x,\eta).$$

Proof of Lemma 2. First of all,

$$\psi(2^{-j}\rho) - \psi(2^{-j+1}\rho) \in \tilde{S}^0_\rho$$

uniformly in j. This follows from Lemma 3 and the fact that the function is supported where $\rho \approx 2^j$. Next,

$$|K_j(x,z)| \leqq |\int a_j(x,\xi)\, e^{-2\pi i x\cdot z}\, d\xi|$$

$$\leq c \int_{B^*(x, c2^j)} d\xi = c\, V_x^*(c2^j) \simeq V_x(2^{-j})^{-1},$$

by Propositions 4 and 5.

This is the case $M = 0$ of Lemma 2a).

Choose η so that $|(z,\eta)| = 1$ and $N_x(z) = 1/\rho(x,\eta)$. Then

$$|K_j(x,z)| = |(z,\eta)^{M'} K_j(x,z)|$$

$$\leqq (2\pi)^m \int |(\eta, \frac{\partial}{\partial \xi})^{M'} a_j(x,\xi)| d\xi$$

$$\leqq C \int_{\rho(x,\xi) \simeq 2^j} \theta\left(\frac{\rho(x,\eta)}{\rho(x,\xi)}\right)^{M'} d\xi$$

$$\leqq C\, V_x(2^{-j})^{-1} \quad (2^j N_x(z))^{-M} .$$

To prove b), notice that $(\lambda, \frac{\partial}{\partial z}) K_j(x,z)$ corresponds to the product of (λ, ξ) with $a_j(x,\xi)$. Proposition 6 implies that

$$|(\lambda,\xi)| \lesssim \theta(N_x(\lambda)\, \rho(x,\xi)),$$

which is exactly the additional factor desired. The proof of c) is similar to b).

<u>Proof of Lemma 1</u>.

$$K(x,z) = \sum_{j=0}^{\infty} K_j(x,z).$$

We will prove Lemma 1 with $K(x,z)$ replaced by any finite sub-sum, although we will continue to denote it $K(x,z)$. ($K(x,z)$ is only defined as a limit of such sums anyway.)

Suppose $N_x(z) = \delta$. Then

$$K(x,z) = \sum_{2^j \delta \leq 1} K_j(x,z) + \sum_{2^j \delta > 1} K_j(x,z). \quad \text{So}$$

$$|K(x,z)| \leq C \sum_{2^j\delta \leq 1} V_x(2^{-j})^{-1} + C_M \sum_{2^j\delta > 1} V_x(2^{-j})^{-1} (2^j\delta)^{-M}.$$

Proposition 5c) shows that the first sum is majorized by a geometric series dominated by its first term $V_x(\delta)^{-1}$. Proposition 5b) shows that the same is true for the second sum for sufficiently large M. The remainder of the estimates of Lemma 1 are deduced from their analogues in Lemma 2 in a similar way.

To prove Theorem 9, for $1 < p \leq 2$, it suffices to verify (ii)$'$ for the kernel $k(x,y) = K(x, x-y)$. For boundedness in the range $2 \leq p < \infty$, we will have to prove a similar inequality for $\widetilde{k}(x,y) = \overline{K}(y, y-x)$, the kernel of the adjoint of T.

Let us begin with the adjoint estimate.

$$\int_{x \in {}^c B(y_0, c\delta)} |K(y, y-x) - K(y_0, y_0-x)| dx$$

$$\leq \int_{x \in {}^c B(y_0, c\delta)} |K(y, y-x) - K(y_0, y-x)| + |K(y_0, y-x) - K(y_0, y_0-x)| dx.$$

The two integrands will be treated in identical fashion, so we will only look at one of them.

Denote $y_t = (1-t)y_0 + ty$. Then

$$|K(y_0, y-x) - K(y_0, y_0-x)| \leq \int_0^1 (y-y_0, \tfrac{\partial}{\partial z}) K(y_0, y_t-x)\, dt.$$

Thus it suffices to show that

$$\int_{x \in {}^c B(y_0, c\delta)} |(y-y_0, \tfrac{\partial}{\partial z}) K(y_0, y_t - x)| dx \leq A \quad \text{for all } t,\ 0 \leq t \leq 1.$$

In fact, the left-hand side is dominated by

$$\int_{x\in {}^cB(y_0, c\delta)} V_{y_0}(y_t - x)^{-1}\, \theta\left(\frac{N_{y_0}(y-y_0)}{N_{y_0}(y_t - x)}\right) dx.^*$$

We can estimate this integral by breaking it into annuli $\{x: N_{y_0}(x) \simeq 2^j\delta\} = A_j$.

For $x \in A_j$, (j sufficiently large)

$$\frac{N_{y_0}(y-y_0)}{N_{y_0}(y_t - x)} \simeq \frac{N_{y_0}(y-y_0)}{N_{y_0}(y_0 - x)} \lesssim \frac{\delta}{2^j\delta} = 2^{-j}, \text{ by Proposition 3.}$$

Furthermore,

$$\int_{A_j} V_{y_0}(y_t - x)^{-1}\, dx \simeq V_{y_0}(y_0 - x)^{-1}\, V_{y_0}(y_0 - x) \leq A.$$

Thus

$$\int_{x\in {}^cB(y_0, c\delta)} V_{y_0}(y_t - x)^{-1}\theta\left(\frac{N_{y_0}(y-y_0)}{N_{y_0}(y_t - x)}\right) dx \lesssim A \sum_j 2^{-j} \lesssim A.$$

For the kernel $k(x,y) = K(x, x-y)$, the proof is slightly harder. A similar argument shows that

$$\int_{x\in {}^cB(y_0, c\delta)} |K(x, x-y) - K(x, x-y_0|dx$$

$$\leq \int_{x\in {}^cB(y_0, c\delta)} V_x(x-y_t)^{-1}\theta\left(\frac{N_x(y_0 - y)}{N_x(x-y_t)}\right) dx.$$

The denominator $N_x(x-y_t) \simeq N_x(x-y_0) \simeq N_{y_0}(x-y_0)$ is the same as the previous one, but the numerator is not. We need Proposition 11:

---

*It will be convenient here to use the abbreviation $V_y(z)$ for $V_y(N_y(z))$.

$$N_x(y_0-y) \lesssim N_{y_0}(y_0-y) + N_x(x-y_0)^{2/3} N_{y_0}(y-y_0)^{1/3}.$$

The first term $N_{y_0}(y_0-y)$ is the same as before. The second term gives rise to a factor in the integrand of

$$\frac{N_x(x-y_0)^{2/3} N_{y_0}(y-y_0)^{1/3}}{N_x(x-y_0)} = \frac{N_{y_0}(y-y_0)^{1/3}}{N_x(x-y_0)^{1/3}},$$

which is just the 1/3 power of the previous factor. In the same way as above, it gives rise to a geometric series with terms like $2^{-j/3}$ instead of 2^{-j}. This concludes the proof.

§7. L^2 estimates

The conditions of the class $\tilde{S}^m_\rho$ are too weak to guarantee L^2 estimates and hence L^p estimates by themselves: They allow too much growth upon differentiation in x. One sort of condition we might be inclined to add is

$$|\nabla_x a(x,\xi)| \lesssim \rho(x,\xi)^m \, \|\xi\| / \rho(x,\xi),$$

and its analogue for higher derivatives. If $a(x,\xi)$ were a symbol of order $m=2$ we would then expect that

$$(16) \qquad |(\eta, \frac{\partial}{\partial \xi}) \nabla_x a(x,\xi)| \lesssim \rho(x,\xi)^2 \frac{\|\xi\|}{\rho(x,\xi)} \left(\frac{\rho(x,\eta)}{\rho(x,\xi)} + \frac{\rho(x,\eta)^2}{\rho(x,\xi)^2} \right).$$

However, this is false for $Q(x,\xi) = \sum_j |L^j_x(\xi)|^2$, and so no interesting application of this type of condition could be expected.

To see this, let us carry out the computation.

$$\frac{1}{2}(\eta, \frac{\partial}{\partial \xi}) \nabla_x Q(x,\xi) = \sum_j [\nabla_x L^j_x(\eta)]\, L^j_x(\xi) + [\nabla_x L^j_x(\xi)]\, L^j_x(\eta).$$

The best bounds we can expect are

$$|\nabla_x L_x^j(\eta)| \lesssim \|\eta\| \quad \text{and} \quad |L_x^j(\xi)| \leq \rho(x,\xi).$$

So the final bound is

$$\|\eta\|\, \rho(x,\xi) + \|\xi\|\, \rho(x,\eta)$$

The term $\|\eta\|\, \rho(x,\xi)$ may be larger than the right-hand side of (16), as when $\rho(x,\eta)^2 \simeq \|\eta\|$ and $\rho(x,\xi) \simeq \|\xi\|$.

We can conclude from this kind of reasoning that conditions bearing only on the size of derivatives of the symbols we are interested in are either too weak to guarantee L^2 boundedness, or too strong in that they would exclude the desired applications. What is the way out of this dilemma? We need to reexamine the basic example given in Proposition 1 of §4. This leads us to our definition of the class S_ρ^m.

Recall that $\hat{S}_\rho^m$ is the class of symbols that satisfies the $\tilde{S}_\rho^m$ conditions in ξ derivatives only. For $\varphi(x) \equiv 1$, the class S_ρ^m is the collection of symbols $a(x,\xi) \in \hat{S}_\rho^m$ satisfying

(17) $$\frac{\partial}{\partial x_j} a(x,\xi) = \sum_{k=1}^{n} a_k(x,\xi)\, \xi_k + a_0(x,\xi)$$

where $a_k \in \hat{S}_\rho^{m-1}$ and $a_0 \in \hat{S}_\rho^m$.

(18) $\frac{\partial}{\partial x_j} a_k$ and $\frac{\partial}{\partial x_i} a_0$ can be written as in (17) (with m replaced by $m-1$ in the case of a_k). The new symbols in the decomposition of $\frac{\partial}{\partial x_i} a_k$ and $\frac{\partial}{\partial x_i} a_0$ can be written as in (17), and so on ad infinitum.

Note that if $a(x,\xi) \in S_\rho^m$, then

(19) $$\frac{\partial}{\partial x_j} a(x,\xi) = \sum_{k=1}^{n} a_k(x,\xi)\,\xi_k + a_0(x,\xi)$$

where $a_k \in S_\rho^{m-1}$ and $a_0 \in S_\rho^m$.

In general, when φ is not necessarily constant, we make a similar definition with condition (17) replaced by

(20) $$(\lambda, \frac{\partial}{\partial x})\, a(x,\xi) = (\lambda, d\varphi(x))\, Q_x^1(\xi)\, \tilde{a}_0(x,\xi) + \sum_{k=1}^{n} a_k(x,\xi)\,\xi_k + a_0(x,\xi),$$

where $\tilde{a}_0 \in \hat{S}_\rho^{m-2}$, $a_k \in \hat{S}_\rho^{m-1}$, $a_0 \in \hat{S}_\rho^m$.

As before, the recursion implies that we can strengthen the conditions in (20) to $\tilde{a}_0 \in S_\rho^{m-2}$, $a_k \in S_\rho^{m-1}$ and $a_0 \in S_\rho^m$.

We would also like to record the iterated form of the condition for membership in S_ρ^m. $S_\rho^m = \left\{ a(x,\xi) \ :\ \prod_{j=1}^{N} (\lambda_j, \frac{\partial}{\partial x})\, a(x,\xi) \right.$ is a sum of terms of the form

$$\Big(\prod_{\beta \in B} (d\varphi(x), \lambda_\beta)\Big)\, Q_x^1(\xi)^{|B|}\, p(x,\xi,\lambda)\, b(x,\xi)$$

where (i) $A \cup B = \{1,\ldots,N\}$ a disjoint union

(ii) $p(x,\xi,\lambda)$ is smooth in x and λ and of degree $d \leq |A|$ in ξ

(iii) $b(x,\xi) \in S_\rho^{m-d-2|B|}$.

For most purposes — in particular the estimates of this section — condition (iii) can be replaced by $b(x,\xi) \in \hat{S}_\rho^{m-d-2|B|}$, which no longer involves any self-reference. One should keep in mind, too, that only finitely many derivatives are necessary for any single estimate.

Proposition 14. $S^m_\rho \subset \tilde{S}^m_\rho$.

We will only give the main ingredients of the proof. For the details see the Appendix.

Case 1. $\varphi(x) \equiv 1$, $m = 0$.

A typical estimate required is

(21) $$|(\eta, \frac{\partial}{\partial \xi})(\lambda, \frac{\partial}{\partial x}) a(x,\xi)| \leq C\,\theta\,(\frac{\rho(x,\eta)}{\rho(x,\xi)})\,\theta\,(\rho(x,\xi)\,N_x(\lambda)).$$

But, $|(\eta, \frac{\partial}{\partial \xi})(\lambda, \frac{\partial}{\partial x}) a(x,\xi)| = |(\eta, \frac{\partial}{\partial \xi})[\sum_{j,k} \lambda_j a^j_k \xi_k) + \sum_j \lambda_j a^j_0]|$

with $a^j_k \in \hat{S}^{-1}_\rho$, $a^j_0 \in \hat{S}^0_\rho$.

This is dominated by

$$\|\lambda\|\,[\,\|\eta\|\rho(x,\xi)^{-1} + \|\xi\|\rho(x,\xi)^{-1}\theta\,(\frac{\rho(x,\eta)}{\rho(x,\xi)}) + \theta\,(\frac{\rho(x,\eta)}{\rho(x,\xi)})].$$

Now trivial estimates $\|\lambda\| \lesssim N_x(\lambda)$, $\|\eta\| \lesssim \rho(x,\eta)^2$ etc. yield (21).

Case 2. φ not necessarily constant.

A typical term that does not appear in Case 1 is

$|(\lambda, d\varphi)|Q^1_x(\xi)\,\theta\,(\frac{\rho(x,\xi)}{\rho(x,\xi)})$. This requires an appropriate bound for $|(\lambda, d\varphi)|$. Proposition 6 implies $|(\lambda, d\varphi)| \lesssim N_x(\lambda)\,|Q(x,d\varphi)|^{1/2} + N_x(\lambda)^2\,\|d\varphi\|$. Also, $\|d\varphi\| \lesssim$ const. and $|Q(x,d\varphi)| \lesssim \varphi(x)^2$ (see (8), §4). Therefore, $|(\lambda, d\varphi)| \lesssim N_x(\lambda)\,\varphi(x) + N_x(\lambda)^2$. Finally,

$$(N_x(\lambda)\,\varphi(x) + N_x(\lambda)^2)\,Q^1_x(\xi)\,\theta\,(\frac{\rho(x,\eta)}{\rho(x,\xi)}) \leqq$$

$$\leqq [N_x(\lambda)\,\rho(x,\xi) + N_x(\lambda)^2\,\rho(x,\xi)^2]\,\theta\,(\frac{\rho(x,\eta)}{\rho(x,\xi)}).$$ qed

Proposition 15. $\rho^m \in S^m_\rho$.

The proof is in the Appendix.

Theorem 10. If $a \in S^0_\rho$, then $a(x, D)$ is a bounded operator on L^2 (and consequently on L^p, $1 < p < \infty$).

Proposition 16. If $\varphi(x) \equiv 1$ and $a(x, \xi) \in S^0_\rho$, then $a(x,\xi) \in S^0_{1/2, 1/2}$, i.e.,

$$(22) \qquad |\partial_\xi^\alpha \partial_x^\beta a(x, \xi)| \le C_{\alpha, \beta}(1 + \|\xi\|)^{-\frac{|\alpha|}{2} + \frac{|\beta|}{2}} .$$

A theorem of Calderón and Vaillancourt [9] says that a symbol in the class $S^0_{1/2, 1/2}$ gives rise to a bounded operator on L^2. Thus in the case $\varphi(x) \equiv 1$, the L^2 part of our theorem is a special case of theirs.

To prove (22), note that $\partial_x^\beta a(x, \xi) = \sum_{|\gamma| \le |\beta|} a_\gamma(x, \xi)\, \xi^\gamma$ where $a_\gamma \in \hat{S}_\rho^{-|\gamma|}$. Now $\partial_\xi^\alpha (a_\gamma(x, \xi)\, \xi^\gamma)$ is made up of terms like

$$(\partial_\xi^{\alpha_1} a_\gamma(x, \xi))\,(\partial_\xi^{\alpha_2} \xi^\gamma) \quad \text{with} \quad \alpha_1 + \alpha_2 = \alpha .$$

And $|\partial_\xi^{\alpha_1} a_\gamma(x, \xi)|\ |\partial_\xi^{\alpha_2} \xi^\gamma| \leqq \rho(x, \xi)^{-|\gamma| - |\alpha_1|} \|\xi\|^{|\gamma| - |\alpha_2|} \lesssim$

$$\lesssim (1 + \|\xi\|)^{-|\alpha_1|/2 + |\gamma|/2 - |\alpha_2|} \lesssim (1 + \|\xi\|)^{|\beta|/2 - |\alpha|/2},$$

because $|\gamma| \le |\beta|$.

The general case. The proof given here is similar to the proof of Theorem 2′. Instead of decomposing the kernel, we decompose the symbol.

Recall that we can write (see the decomposition preceding Lemma 2)

$$a(x,\xi) = \sum a_j(x,\xi)$$

with
$$a_j(x,\xi) \simeq \begin{cases} a(x,\xi) & \rho(x,\xi) \simeq 2^j \\ 0 & \text{otherwise.} \end{cases}$$

Define $T_j = a_j(x,D)$. We wish to apply the quasi-orthogonality lemma of §2 of Chapter I.

We shall prove that

(23) $$\|T_j^* T_k\| \leqq A\, 2^{-|j-k|},$$

(24) $$\|T_j T_k^*\| \leq A\, 2^{-|j-k|}.$$

Recall that $T_j f(x) = \int K_j(x, x-z) f(z) dz$ where K_j satisfies the estimates of Lemma 2, Section 6. (The proof only depended on $\tilde{S}^0_\rho$ conditions on the symbol $a(x,\xi)$.) We will also need an extra ingredient that uses S^0_ρ conditions, not just $\tilde{S}^0_\rho$:

<u>Main lemma</u>

(25) $$\left|\int K_j(z, x-z) dz\right| \leqq A\, 2^{-j}, \quad j=0,1,2,\ldots$$

Before proving the lemma let us remark that an even stronger statement is true for the adjoint.

(26) $$\int K_j(x, x-z) dz = 0 \qquad j=1,2,\ldots$$

because $a_j(x,\xi)$ vanishes for ξ near 0 when $j > 0$.

<u>Proof of main lemma.</u>* The special case $\varphi(x) \equiv 1$.

$$\int K_j(z, x-z) dz = \iint e^{i(z-x)\cdot\xi} a_j(z,\xi)\, d\xi\, dz$$

$$= \int e^{-ix\cdot\xi}\left\{\int e^{iz\cdot\xi} a_j(z,\xi)\, dz\right\} d\xi.$$

*For simplicity of notation in the calculations below we omit the factors 2π.

$\int e^{iz\cdot\xi} a_j(z,\xi)\,dz$ is supported (approximately) in

$\{\xi : 2^{j/2} < \|\xi\| < 2^j\}$, so it certainly suffices to prove

(27) $$\left|\int e^{iz\cdot\xi} a_j(z,\xi)\,dz\right| \leqq C_N \|\xi\|^{-N}.$$

Replace $e^{iz\cdot\xi}$ by $\|\xi\|^{-2N}\Delta_z^N e^{iz\cdot\xi}$ and integrate by parts.

$$\left|\int e^{iz\cdot\xi} a_j(z,\xi)\,dz\right| \le \|\xi\|^{-2N}\int\left|\Delta_z^N a_j(z,\xi)\right|dz.$$

We just proved that $|\Delta_z^N a_j(z,\xi)| \lesssim \|\xi\|^N$ (Proposition 16). In the balance, we gain a power $\|\xi\|^{-N}$ and (27) follows.

<u>The general case</u>

Once again we will prove (27). The idea is to integrate by parts as above, but in selective directions. For clarity, we will first carry out the case related to the Kannai example.

$$Q(x,\xi) = x_1(\xi_2^2 + \dots + \xi_n^2)$$

$$x = (x_1,\dots,x_n),\quad \xi = (\xi_1,\dots,\xi_n),\quad \xi' = (\xi_2,\dots,\xi_n).$$

$$\rho(x,\xi) = (Q(x,\xi)^2 + \|\xi\|^2)^{1/4}.$$

<u>Case 1</u>. $\|\xi'\| > \|\xi\|^{3/4}$

Denote $\Delta_{z'} = \dfrac{\partial^2}{\partial x_2^2} + \dots + \dfrac{\partial^2}{\partial x_n^2}$.

Recall that for $a \in S_\rho^0$,

(28) $$(\lambda, \frac{\partial}{\partial x})\,a(x,\xi) = (\lambda, d\varphi)(\xi_2^2 + \dots + \xi_n^2)\,a_{-2}(x,\xi) + \sum_j a_{-1}^j \xi_j + a_0.$$

The factor $(\lambda, d\varphi)$ vanishes when $\lambda_1 = 0$, because $d\varphi = (1, 0, \ldots, 0)$. Therefore

$$|(1 - \Delta_{z'})^N a_j(z, \xi)| \lesssim \|\xi\|^{2N} \rho^{-2N} \lesssim \|\xi\|^N.$$

Making use of the substitution $(1 - \Delta_{z'})^N e^{iz\cdot\xi} = (1 + \|\xi'\|^2)^N e^{iz\cdot\xi}$:

$$\left|\int a_i(z, \xi) e^{iz\cdot\xi} dz\right| = \frac{1}{(1 + \|\xi'\|^2)^N} \left|\int \left[(1 - \Delta_{z'})^N a_j(z, \xi)\right] e^{iz\cdot\xi} dz\right|$$

$$\lesssim \frac{\|\xi\|^N}{\|\xi'\|^{2N}} < \frac{\|\xi\|^N}{\|\xi\|^{3N/2}} = \|\xi\|^{-N/2}.$$

<u>Case 2</u>. $\quad \|\xi'\| \leq \|\xi\|^{3/4}$.

This time we use the full Laplacian.

$$\int a_j(z, \xi) e^{iz\cdot\xi} dz = \frac{1}{(1 + \|\xi\|^2)^N} \int \left[(1 - \Delta_z)^N a(z, \xi)\right] e^{iz\cdot\xi} dz.$$

From (28) we get

$$|(1 - \Delta_z)^N a(x, \xi)| \lesssim \|\xi'\|^{4N} \rho^{-4N} + \text{intermediate terms} + \|\xi\|^{2N} \rho^{-2N}.$$

The intermediate terms are dominated by the outer ones, so in all

$$\left|\int a_j(z, \xi) e^{iz\cdot\xi} dz\right| \lesssim \|\xi\|^{-2N} (\|\xi'\|^{4N} \rho^{-4N} + \|\xi\|^{2N} \rho^{-2N})$$

$$\lesssim \|\xi\|^{-2N} (\|\xi'\|^{4N} \|\xi\|^{-2N} + \|\xi\|^N)$$

$$\lesssim \|\xi\|^{-2N} (\|\xi\|^{3N} \|\xi\|^{-2N} + \|\xi\|^N) \lesssim \|\xi\|^{-N}.$$

We now turn to the general case. The argument is somewhat more elaborate.

We may assume that $\|\xi\| \geq 1$, and that we have restricted our attention to a sufficiently small neighborhood of the submanifold where $\varphi(x) = 0$ (i.e., $d\varphi \neq 0$ there).

We let π_x^1 be the orthogonal projection in the direction of $(d\varphi)(x)$, and π_x^2 the projection on the orthogonal complement. We write

$$\xi = \bar{\xi} + \xi' = \pi_x^1(\xi) + \pi_x^2(\xi),$$

and define the vector λ by

$$\lambda = \bar{\xi}\,\|\xi\|^{-1/4} + \xi'.$$

Note that λ is a function of x and ξ, but satisfies the inequalities:

$$(*)\qquad \begin{aligned} &|\partial_x^\alpha \lambda| \leq c_\alpha \|\xi\|, \qquad \text{all } \alpha \\ &|(\lambda,\xi)| \geq \|\xi\|^{2-1/4}, \qquad |(\lambda,\xi)| \geq \|\xi'\|^2. \end{aligned}$$

Let D be the first-order differential operator given by

$$Du = \frac{1}{i(\lambda,\xi)}(\lambda\cdot\nabla_x)u$$

Note that $De^{ix\cdot\xi} = e^{ix\cdot\xi}$, and so if D^t denotes the transpose of D, then since a has compact support in x

$$\int e^{ix\cdot\xi} a(x,\xi)\,dx = \int e^{ix\cdot\xi}(D^t)^N a(x,\xi)\,dx$$

for each integer $N \geq 0$. We claim that in fact

$$|(D^t)^N a(x,\xi)| \leq C_N \|\xi\|^{-N/4},$$

which will prove our lemma.

To see this, consider the various terms that occur in the N-fold differentiation. If the N differentiations fall entirely on λ and its derivatives, then because the inequalities (*) we get a gain of $\|\xi\|^{-3/4\,N}$, which is better than what we need. On the other hand the derivatives which involve some differentiation of λ and some of a are controlled (in view of what we have said about the λ derivatives) by the derivatives which involve only a. Consider for example the first derivative. Because of (28) this gives us a factor of the order of magnitude

$$\frac{|(\lambda, d(\varphi)|Q^1_x(\xi)}{\rho^2\,|\lambda\cdot\xi|} + \frac{\|\lambda\|\ \|\xi\|}{\rho|\lambda\cdot\xi|}\,.$$

By the inequalities (*) the second term is dominated by

$$\frac{\|\xi\|\ \|\xi\|}{\rho\|\xi\|^{2-1/4}} = \frac{\|\xi\|^{1/4}}{\rho} \leq \|\xi\|^{-1/4}\,.$$

To handle the first term we have that

$$Q^1_x(\xi) \leq 2\,Q^1_x(\bar{\xi}) + 2\,Q^1_x(\xi'), \quad \text{since } \xi = \bar{\xi} + \xi'.$$

But $Q^1_x(\bar{\xi}) \leq$ constant $\rho(x,\bar{\xi})\ \|\bar{\xi}\|$, since $\bar{\xi}$ is in the direction of $(d\varphi)(\lambda)$, and $Q^1_x(d\varphi) = 0$ when $\varphi = 0$ (see (8) in the proof of Proposition 7). This again leads to contribution $\leq C\|\xi\|^{-1/4}$. Finally we must handle the term

$$\frac{|(\lambda\ , d\varphi)|\ Q^1_x(\xi')}{\rho^2|\lambda\cdot\xi|}\,.$$

Now $\quad |\lambda\ , d\varphi| \leq C\|\bar{\xi}\|\ \ \|\xi\|^{-1/4}$

and $\quad Q^1_x(\xi') \leq C\|\xi'\|^2$. Also by (*), $|\lambda\ , \xi| \geq \|\xi'\|^2$. Thus this term is dominated by

$$\frac{C\|\bar{\xi}\| \; \|\xi\|^{-1/4}}{\rho^2} .$$

From this we can conclude that

$$|(D^t)^N a(x,\xi)| \leq C_N \|\xi\|^{-N/4}$$

and the lemma is proved.

Conclusion of the proof of Theorem 10

We will make use of the following well-known fact. If $\int |L(x,y)| dx \leq A$ and $\int |L(x,y)| dy \leq A$, then $Tf(x) = \int L(x,y) f(y) dy$ is a bounded operator on L^2 with norm $\leq A$.

Recall from Lemma 2 that

(29) $\quad |K_i(x,z)| \leq C_M V_x(2^{-j})^{-1} (2^j N_x(z))^{-M}$. Therefore,

$$\int |K_i(x, x-z)| dz \lesssim V_x(2^{-j})^{-1} \int \min(1, 2^j N_x(x-z))^{-M}$$

$$\lesssim V_x(2^{-j})^{-1} \Bigg(\int_{N_x(x-z) \leq 2^{-j}} dz + \sum_{k=0}^{\infty} \int_{2^{-j+k} < N_x(x-z) \leq 2^{-j+k+1}} (2^j N_x(x-z))^{-M} dz \Bigg).$$

For sufficiently large M the final sum is dominated by a geometric series. Therefore,

$$\int |K_j(x, x-z)| dz \leq A.$$

A slightly more complicated argument gives

$$\int |K_j(z, x-z)| dz \leq A.$$

(The extra argument needed was already used in the proof of (ii)$'$.) It follows that

$$\|T_j\| \leq A \text{ and } \|T_j^*\| \leq A.$$

Next we look at the kernels of $T_j^* T_\ell$ and $T_j T_\ell^*$.

(a) $$K_{j,\ell}(x,y) = \int K_j(z, x-z)\, \overline{K}_\ell(z, z-y)\, dz$$

(b) $$\int K_j(x, x-z)\, \overline{K}_\ell(y, z-y)\, dz$$

We begin with a) in the case $j \geq \ell + c$. Estimate (29) says that K_j is essentially concentrated on the ball in $x-z$ of radius 2^{-j}. We will therefore try to replace z by x in $\overline{K}_\ell$.

$$K_{j,\ell}(x,y) = \int K_j(z, x-z)\, \overline{K}_\ell(x, x-y)\, dz$$

$$+ \int K_j(z, x-z)\, (\overline{K}_\ell(z, z-y) - \overline{K}_\ell(x, x-y))\, dz$$

The first term can be written as

$$\overline{K}_\ell(x, x-y) \int K_j(z, x-z)\, dz .$$

We just showed that the kernel $\overline{K}_\ell(z, z-y)$ operates on L^2 with norm $\leq A$. The integral $\int K_j(z, x-z)\, dz$ is small by our main lemma.

The main part of the second term arises when $N_x(x-z) \lesssim 2^{-j}$. (We will make some remarks about the rest later.) The estimates from Lemma 2

$$|(\lambda, \partial/\partial z) K_\ell(x,z)| \lesssim V_x(2^{-\ell})^{-1}\, \theta(2^\ell N_x(\lambda))$$

$$|(\lambda, \partial/\partial x) K_\ell(x,z)| \lesssim V_x(2^{-\ell})^{-1}\, \theta(2^\ell N_x(\lambda))$$

with $\lambda = (x-z)$ imply that heuristically

$$\int \overline{K}_\ell(z, z-y) - \overline{K}_\ell(x, x-y)| \lesssim \sup_{x_t} V_{x_t}(2^\ell)^{-1}\, \theta(2^\ell N_{x_t}(x-z))$$

where x_t is between x and z. Combining this with $|K_j(z,x-z)| \lesssim V_z(2^{-j})^{-1}$, we obtain as a bound for the main part of the second term:

$$\sup_t \int_{N_x(x-z) \lesssim 2^{-j}} V_z(2^{-j})^{-1} V_{x_t}(2^{-\ell})^{-1} (2^{\ell} N_x(x-z) + 2^{2\ell} N_x(x-z)^2)\, dz$$

$$\lesssim 2^{\ell-j} \int_{N_x(x-z) \lesssim 2^{-j}} V_x(2^{-j})^{-1} V_x(2^{-\ell})^{-1} dz$$

$$\lesssim 2^{\ell-j} V_x(2^{-\ell})^{-1}.$$

In order to estimate the L^2 operator norm of $T_j^* T_\ell$, we estimate the L^1 norm of the kernel in each variable. Integrating in y with x fixed, the main part of $\overline{K}_\ell(x, x-y)$ is concentrated where $N_x(x-y) \lesssim 2^{-\ell}$, a ball of volume $V_x(2^{-\ell})$. Thus the main term is $2^{\ell-j} V_x(2^{-\ell})^{-1} V_x(2^{-\ell}) \simeq 2^{\ell-j}$. The remaining terms are estimated in terms of integrals with additional convergence factors like $(2^j N_x(x-z))^{-M}$ and $(2^\ell N_x(x-y))^{-M}$, to control the sum when x and z are farther apart than 2^{-j} or x and y are farther apart than $2^{-\ell}$.

If y is fixed and we wish to integrate in x, the main term still comes from the region where $N_x(x-y) \simeq N_y(x-y) \lesssim 2^{-\ell}$. Thus the main term is the same. We use Proposition 11 to show that the convergence factors like $(2^\ell N_x(x-y))^{-M}$ tends to zero geometrically as x is further from y (i.e., as $N_y(x-y)$ increases from $2^{-\ell}$ to 1). This concludes the presentation of the heuristics for the proof of

Theorem 10. The actual details are more involved and they are as follows:

In view of (25) and (26) it suffices to give estimates for the kernels

$$\int K_j(z, x-z)\overline{K}_\ell(z, z-y)\,dz = \int K_j(z, x-z)[\overline{K}_\ell(z, z-y) - \overline{K}_\ell(x, x-y)]\,dz,$$

and

$$\int K_j(x, x-z)\overline{K}_\ell(y, z-y)\,dz = \int K_j(x, x-z)[\overline{K}_\ell(y, z-y) - \overline{K}_\ell(y, x-y)]\,dz.$$

Next we write $\overline{K}_\ell(z, z-y) - \overline{K}_\ell(x, x-y)$ as

$$\int_0^1 \Big(\langle \nabla_1 \overline{K}_\ell(x_t, z-y), z-x\rangle + \langle \nabla_2 \overline{K}_\ell(x, x_t-y), z-x\rangle \Big)\,dt$$

with $x_t = tx + (1-t)z$. A similar integration is used for the second kernel. This reduces matters to the estimates of three integrals (uniformly in t).

I $\quad \int K_j(x, x-z)\langle \nabla_2 \overline{K}_\ell(y, x_t-y), z-x\rangle\,dz$

II $\quad \int K_j(z, x-z)\langle \nabla_1 \overline{K}_\ell(x_t, z-y), z-x\rangle\,dz$

III $\quad \int K_j(z, x-z)\langle \nabla_2 \overline{K}_\ell(x, x_t-y), z-x\rangle\,dz$

We do II first. Estimating according to Lemma 2, we get

$$\int V_z(2^{-j})^{-1} C_N[2^j N_z(z-x)]^{-N} V_{x_t}(2^{-\ell})^{-1} C_M[2^\ell N_{x_t}(z-y)]^{-M}[2^\ell N_{x_t}(z-x)]^p\,dz$$

$$N, M \text{ arbitrary} \geq 0, \quad p \geq 1.$$

We use $\quad N_{x_t}(z-x) \approx N_x(z-x) \approx N_z(z-x)$,

also $\quad V_{x_t}(2^{-\ell})^{-1} \lesssim V_z(2^{-\ell})[1 + \left[\frac{N_z(z-x)}{2^{-\ell}}\right]^m \lesssim V_z(2^{-\ell})^{-1}[2^\ell N_z(z-x)]^m, \quad m \geq 0$

Finally, $N_z(z-y) \lesssim N_{x_t}(z-y) + N_{x_t}(z-y)^{1/3} N_x(z-x)^{2/3}$

Hence, either $N_{x_t}(z-y) \gtrsim N_z(z-y)$

or $N_{x_t}(z-y) \gtrsim N_z(z-y)^3 N_x(z-x)^{-2}$

Thus we get two possibilities:

$$C_N C_M \int V_z(2^{-j})^{-1} V_z(2^{-\ell})^{-1} [2^j N_z(z-x)]^{-N} [2^\ell N_z(z-x)]^m \times$$
$$\times [2^\ell N_z(z-y)]^{-M} [2^\ell N_z(z-x)]^p \, dz$$

or

$$C_N C_M \int V_z(2^{-j})^{-1} V_z(2^{-\ell})^{-1} [2^j N_z(z-x)]^{-N} [2^\ell N_z(z-x)]^m [2^\ell N_z(z-y)^3 N_x(z-x)^{-2}]^{-M} \times$$
$$\times [2^\ell N_z(z-x)]^p \, dz$$

$$= C_N C_M \int V_z(2^{-j})^{-1} V_z(2^{-\ell})^{-1} [2^j N_z(z-x)]^{-N+2M} [2^\ell N_z(z-y)^3]^{-M}$$
$$[2^\ell N_z(z-x)]^{p+m} \cdot 2^{-2Mj} \quad dz$$

$$= C_N C_M \, 2^{2M(\ell-j)} \int V_z(2^{-j})^{-1} V_z(2^{-j})^{-1} [2^\ell N_z(z-x)]^{-N+2M} [2^\ell N_z(z-y)]^{-3M}$$
$$[2^\ell N_z(z-x)]^{p+m} \quad dz$$

and this shows that we get the right estimates when integrating with respect to x or y. E.g., consider what happens when we integrate in x:

$$\int V_z(2^{-j})^{-1} V_z(2^{-\ell})^{-1} C_M [2^\ell N_z(z-y)]^{-3M} \int C_N [2^j N_z(z-x)]^{-N+2M} [2^\ell N_z(z-x)]^{p+m} dx\,dz$$

(but the inner integral $\lesssim 2^{(\ell-j)(p+m)} V_z(2^{-j})$), therefore we have the estimate

$$\lesssim \int V_z(2^{-\ell})^{-1} C_M[2^\ell N_z(z-y)]^{-3M}\, dz \qquad 2^{(\ell-j)(p+m)}$$

$$\lesssim V_y(2^{-\ell})^{-1} \int [1+\left[\frac{N_z(z-y)}{2^{-\ell}}\right]^M] C_M[2^\ell N_z(z-t)]^{-3M}\, dz \qquad 2^{(\ell-j)(p+m)}$$

$$\lesssim 2^{\ell-j} \qquad \text{since} \quad p \geq 1.$$

This does II. I and III are similar. We do I; estimating it, we get:

$$\int V_x(2^{-j})^{-1} C_N[2^j N_x(x-z)]^{-N} V_y(2^{-\ell})^{-1} C_M[2^\ell N_y(y-x_t)]^{-M} [2^\ell N_y(z-x)]^p\, dz$$

$$p \geq 1 \qquad N, M \geq 0.$$

Now $N_y(z-x) \leq N_x(z-x) + N_x(z-x)^{1/3} N_x(y-x)^{2/3}$. Hence this integral is also dominated by:

$$\int V_x(2^{-j})^{-1}\, 2^{(\ell-j)p}\, C_N[2^j N_x(x-z)]^{-N+p} V_y(2^{-\ell})^{-1} C_M[2^\ell N_y(y-x_t)]^{-M}\, dz \, +$$

$$\int V_x(2^{-j})^{-1}\, 2^{(\ell-j)p/3}\, C_N[2^j N_x(x-z)]^{-N+p/3}\, [2^\ell N_x(x-y)]^{2/3\,p}\, V_y(2^{-\ell})^{-1} \times$$

$$\times\, C_M[2^\ell N_y(y-x_t)]^{-M}\, dz$$

Now let $R(x,y,t) = \{z \mid N_y(y-x_t) \leq c\, N_y(y-x)\}$. This is the bad region, for on $\mathbb{R}^n - R(x,y,t)$, $cN_y(y-x) \leq N_y(y-x_t)$ and the integral is dominated by

$$\int V_x(2^{-j})^{-1} C_N[2^j N_x(x-z)]^{-N} V_y(2^{-\ell})^{-1} C_M[2^\ell N_y(y-x)]^{-M} \times$$

$$\times\, [2^\ell N_x(z-x) + 2^\ell N_x(z-x)^{1/3} N_x(y-x)^{2/3}]^p \quad dz.$$

$$\lesssim V_x(2^{-j})^{-1} V_y(2^{-\ell})^{-1} C_M[2^\ell N_y(y-x)]^{-M} \int C_N[2^j N_x(x-z)]^{-N} \times$$

$$\times [2^{\ell} N_x(z-x) + 2^{\ell/3} N_x(z-x)^{1/3} \cdot 2^{2/3\,\ell} N_x(y-x)^{2/3}]^p =$$

$$V_x(2^{-j})^{-1} V_y(2^{-\ell})^{-1} C_M [2^{\ell} N_y(y-x)]^{-M} V_x [2^{(\ell-j)P} + 2^{(\ell-j)^{p/3}} [2^{\ell} N_x(y-x)]^{2p/3}].$$

Hence, integrating in x or y gives an estimate $2^{(\ell-j)P} \lesssim 2^{\ell-j}$. Thus we only have to worry about the integral over $R(x,y,t)$, i.e., we now assume

$$N_y(y-x_t) \leq c\, N_y(y-x).$$

Now write $\quad x_t = x + t(z-x)$

and define z_0 by $y = x + t(z_0 - x)$, $z_0 = \frac{1}{t}(y-x) + x$.

Then

$$y - x_t = t(z_0 - z)$$

$$x_t - x = t(z-x)$$

$$y - x = t(z_0 - x)$$

$$y - z_0 = y - x - \frac{1}{t}(y-x) = (1 - \frac{1}{t})(y-x) = \frac{t-1}{t}(y-x)$$

$$\therefore\ y - z_0 = (t-1)(z_0 - x).$$

Now

$$N_y(y-x) = N_y(y - x_t + x_t - x) \lesssim N_y(y-x_t) + N_y(x_t - x)$$

$$\lesssim c\, N_y(y-x) + N_y(x_t - x).$$

$$\therefore\quad N_y(y-x) \lesssim N_y(x_t - x) = N_y(t(z-x)) \leq \sqrt{t}\, N_y(z-x)$$

$$\lesssim \sqrt{t}\, [N_x(z-x) + N_x(z-x)^{1/3} N_x(y-x)^{2/3}].$$

$$\therefore\quad \text{either}\ N_y(y-x) \lesssim \sqrt{t}\, N_x(z-x)$$

$$\text{or}\quad N_y(y-x) \leq \sqrt{t}\, N_x(z-x)^{1/3} N_y(y-x)^{2/3}.$$

$$\therefore\quad N_y(y-x) \lesssim \sqrt{t}\, N_x(z-x).$$

Also $\qquad N_y(y-x_t) = N_y(t(z-z_0)) \geq t\, N_y(z-z_0)$.

But
$$N_{z_0}(z-z_0) \leq N_y(z-z_0) + N_y(z-z_0)^{1/3} N_y(y-z_0)^{2/3}$$
$$\leq N_y(z-z_0) + N_y(z-z_0)^{1/3} N_y(x-z_0)^{2/3}$$
$$\leq N_y(z-z_0) + N_y(z-z_0)^{1/3} N_y(\tfrac{1}{t}(y-x))^{2/3}$$
$$\leq N_y(z-z_0) + \frac{1}{t^{1/3}} N_y(z-z_0)^{1/3} N_y(y-x)^{2/3}.$$

$\therefore$ either $\qquad N_y(z-z_0) \gtrsim N_{z_0}(z-z_0)$

or $\qquad N_y(z-z_0) \gtrsim t\, N_{z_0}(z-z_0)^3 N_y(y-x)^{-2}$.

Hence
$$N_y(y-x_t) \gtrsim \begin{cases} t\, N_{z_0}(z-z_0) \\ t^2 N_{z_0}(z-z_0)^3 N_y(y-x)^{-2} \end{cases}.$$

Thus in the region $R(x,y,t)$ integral I is dominated by

$$\int_R V_x(2^{-j})^{-1} C_N[2^j t^{-1/2} N_y(y-x)]^{-N+p} V_y(2^{-\ell})^{-1} \times$$
$$\times\, C_M[2^\ell t\, N_{z_0}(z-z_0)]^{-M} dz \qquad 2^{(\ell-j)p}$$

or
$$2^{(\ell-j)\frac{p}{3}} [2^\ell N_x(x-y)]^{\frac{2}{3}p} \int V_x(2^{-j})^{-1} C_M[2^j t^{-1/2} N_y(y-x)]^{-N+\frac{p}{3}} \times$$
$$V_y(2^{-\ell})^{-1} \times C_N[2^\ell t\, N_{z_0}(z-z_0)]^{-M} dz$$

or
$$2^{(\ell-j)} \int V_x(2^{-j})^{-1} C_N[2^j t^{-1/2} N_y(y-x)]^{-N+p} V_y(2^{-\ell})^{-1} \times$$
$$\times\, C_M[2^\ell t^2 N_{z_0}(z-z_0)^3 N_y(y-x)^{-2}]^{-M} dz$$

or

$$2^{(\ell-j)\frac{p}{3}}[2^{\ell}N_x(x-y)]^{\frac{2}{3}p}\int V_x(2^{-j})^{-1}C_N[2^j t^{-1/2}N_y(y-x)]^{-N+\frac{p}{3}} \times$$

$$\times V_y(2^{-\ell})^{-1}C_M[2^{\ell}t^2 N_{z_0}(z-z_0)^3 N_y(y-x)^{-2}]^{-M}dz.$$

We can evaluate these to get:

$$2^{(\ell-j)p}V_x(2^{-j})^{-1}V_y(2^{-\ell})^{-1}t^{\frac{N}{2}-\frac{p}{2}-M}C_N[2^jN_x(x-y)]^{-N+p}V_z(2^{-\ell})$$

or

$$2^{(\ell-j)\frac{p}{3}}[2^{\ell}N_y(x-y)]^{\frac{2p}{3}}V_x(2^{-j})^{-1}V_y(2^{-\ell})^{-1}t^{\frac{N}{2}-\frac{p}{6}-M}C_N[2^jN_x(x-y)]^{-N+\frac{p}{3}}V_z(2^{-\ell})$$

or

$$2^{(\ell-j)p}V_x(2^{-j})^{-1}V_y(2^{-\ell})^{-1}t^{\frac{N}{2}-\frac{p}{2}-2M}C_N[2^jN_x(x-y)]^{-N+p}[2^jN_x(x-y)]^{2M} \times$$

$$\times 2^{(\ell-j)2M}V_z(2^{-\ell})$$

or

$$2^{(\ell-j)\frac{p}{3}}[2^{\ell}N_x(x-y)]^{\frac{2}{3}}V_x(2^{-j})^{-1}V_y(2^{-\ell})^{-1}t^{\frac{N}{2}-\frac{p}{6}-2M}C_N[2^jN_x(x-y)]^{-N+\frac{p}{3}} \times$$

$$\times [2^jN_x(x-y)]^{2M}\; 2^{(\ell-j)2M}V_z(2^{-\ell}).$$

In all of these, when we integrate in x or y, with $N_x(x-y) \geq 2^{-j}$, we get the right estimate (since we can take N so large that exponent of t is positive).

Finally, assume $N_x(x-y) \leq 2^{-j}$. Then

$$N_y(y-x_t) \leq c\,N_y(y-x) \leq 2^{-j}.$$

In estimating I, we take $M = 0$ and get

$$\int_R V_x(2^{-j})^{-1} C_N[2^{-j}N_x(x-z)]^{-N} V_y(2^{-\ell})^{-1}[2^{\ell}N_x(x-z)+2^{\ell}N_x(x-z)^{1/3}N_x(x-y)^{2/3}]^p\,dz$$

$$\leq V_y(2^{-\ell})^{-1} \quad 2^{(\ell-j)p}.$$

But $\displaystyle\int_{N_x(x-y)\leq 2^{-j}} 2^{\ell-j} V_y(2^{-\ell})^{-1} \begin{matrix} dx \\ dy \end{matrix} \lesssim$

$$\begin{cases} 2^{\ell-j} V_y(2^{-\ell})^{-1} \, V_y(2^{-j}) \lesssim 2^{\ell-j} \\[2ex] \displaystyle\int_{N_y(x-y)\leq 2^{-j}} 2^{\ell-j} V_x(2^{-\ell})[1+[N_y(x-y)2^{\ell}]^M]dy \lesssim 2^{\ell-j}. \end{cases}$$

This completes the estimates for I. We now deal with III. Estimating this we get

$$\int V_z(2^{-j})^{-1} C_N[2^{j}N_z(z-x)]^{-N} V_x(2^{-\ell})^{-1} C_M[2^{\ell}N_x(y-x_t)]^{-M}[2^{\ell}N_x(z-x)]^p\,dz$$

$$p \geq 1, \quad N, M > 0.$$

Now $\quad V_z(2^{-j})^{-1} \leq V_x(2^{-j})^{-1}\left[1+\left[\dfrac{N_x(z-x)}{2^{-j}}\right]^m\right].$

$$N_z(z-x) = N_x(z-x)$$

$$V_x(2^{-\ell})^{-1} \leq V_y(2^{-\ell})^{-1}\left[1+[2^{\ell}N_x(x-y)]^m\right].$$

$$N_y(y-x_t) \leq N_x(y-x) + N_x(y-x)^{1/3} N_x(x-y)^{2/3}.$$

$\therefore$ either $\quad N_x(y-x_t) \geq N_y(y-x_t)$

or $\quad N_x(y-x_t) \geq N_y(y-x_t)^3 \, N_x(y-x)^{-2}$

$N_x(z-x) \leq N_y(z-x) + N_y(z-x)^{1/3} N_x(x-y)^{2/3}$. Hence the integral

When $z = -2\ell - m + iy$, $a_z \in S_\rho^{-2\ell}$. Case 3 implies that $a_z(x,D)$ is bounded $L_k^p \to L_{k+\ell}^p$. Similarly, when $z = -2\ell - 2 - m + iy$, $a_z \in S_\rho^{-2\ell-2}$ and $a_z(x,D)$ is bounded $L_k^p \to L_{k+\ell+1}^p$. Moreover, since estimates depend on only finitely many derivatives, the norm has polynomial growth as $y \to \pm\infty$. Now the interpolation cited above shows that $a(x,D)^* = a_0(x,D)$ is bounded $L_k^p \to L_{k-m/2}^p$. qed.

Remark. If we appeal to the duality of L_k^p spaces and Proposition 4 below, we can extend the result to all real values of k.

The Lipschitz space Λ_α for $0 < \alpha < 1$ is defined as $\Lambda_\alpha = \{f\colon f$ is continuous, bounded and there exists A s.t. $\|f\|_{L^\infty} \le A$, $\|f(x+t) - f(x)\|_{L^\infty(dx)} \le A|t|^\alpha\}$. The Λ_α norm is the smallest value of A that can be assigned to f.

The definition of Λ_α given above is not the right one for $\alpha = 1$ because it would say that f and its first derivatives are in L^∞. Instead, for $\alpha = 1$ we require $\|f\|_{L^\infty} \le A$ and $\|f(x+t)+f(x-t)-2f(x)\|_{L^\infty(dx)} \le A|t|$. Now for $\alpha > 1$, we say that $f \in \Lambda_\alpha$ if and only if $f \in \Lambda_{\alpha-1}$ and $\partial f/\partial x_j \in \Lambda_{\alpha-1}$ for all j. (See Stein [41], Chapter 5.)

Theorem 12. *If $a(x,\xi) \in S_\rho^m$ and $m < 0$, then $a(x,D)$ is bounded $\Lambda_\alpha \to \Lambda_{\alpha-m/2}$.*

What lies behind the proof of Theorem 12 is another approach to Λ_α by means of the real method of interpolation. Suppose $B_1 \subset B_0$ is a continuous inclusion of Banach spaces. We can define an intermediate space $B_\theta = [B_0, B_1]_\theta$, $0 < \theta < 1$, by $\{f \in B_0\colon \exists A\ \forall t \in (0,1]\ \exists f_0 \in B_0, f_1 \in B_1$

* We will comment on the case $m=0$ later.

for III is dominated by

$$\int V_x(2^{-j})^{-1} C_N[2^j N_x(z-x)]^{-N+m}\, V_y(2^{-\ell})^{-1}\, [2^\ell N_x(x-y)]^m \times$$

$$\times \left[C_M[2^\ell N_y(y-x_t)]^{-M} + C_M[2^\ell N_y(y-x_t)^3 N_x(y-x)^{-2}]^{-M}\right] \times$$

$$\times \left[2^\ell N_y(z-x) + 2^{\ell/3} N_y(z-x)^{1/3}\,[2^{2\ell/3} N_x(y-x)^{2/3}]\right]^p$$

$$\lesssim [2^\ell N_x(x-y)]^{m+\alpha}\,[2^j N_x(y-x)]^{2M}\, 2^{(\ell-j)2M}$$

$$\int V_x(2^{-j})^{-1} C_N[2^j N_x(z-x)]^{-N+m}\, V_y(2^{-\ell})^{-1}$$

$$C_M[2^\ell N_y(y-x_t)]^{-M}[2^\ell N_y(z-x)]^{1-\alpha}\, dz.$$

and this is of the same form as I.

This concludes the proof of Theorem 10.

Chapter III. Further Regularity Theorems and Composition of Operators

§8. Sobolev and Lipschitz spaces

Let L_k^p denote the usual Sobolev space on $\mathbb{R}^n$, $1 < p < \infty$, k real. (See e.g., Stein [41], Chapter 5.)

Theorem 11. If $a \in S_\rho^m$, $m \leq 0$, then $a(x,D)$ is bounded, $L_k^p \to L_{k-m/2}^p$, $k \geq 0$

Lemma 1. Suppose $p(x,\xi)$ is a polynomial in ξ of degree $\leq k$ with smooth coefficients with compact support in x. Then $p(x,\xi) \in S_\rho^{2k}$.

Proof. The product of a symbol of order m_1 and a symbol of order m_2 is a symbol of order $m_1 + m_2$. A function $a(x) \in C_0^\infty$ is clearly a symbol in S_ρ^0. Therefore, it suffices to show that $\xi_j \in S_\rho^2$. In fact,

$$\left|(\eta, \frac{\partial}{\partial \xi})\xi_j\right| = |\eta_j| \leq \rho(x,\eta)^2 \leqq \rho(x,\xi)^2 \left(\frac{\rho(x,\eta)}{\rho(x,\xi)} + \frac{\rho(x,\eta)^2}{\rho(x,\xi)^2}\right).$$

Lemma 2. If $a(x,\xi) \in S_\rho^m$ and $p(x,\xi)$ has degree k as in Lemma 1, then

(1) $$p(x,D) \circ a(x,D) = \sum_j a_j(x,D) \circ p_j(x,D),$$

where $a_j \in S_\rho^m$ and p_j is a polynomial of degree $\leq k$.

Proof. A typical case is $p(x,D) = \dfrac{\partial}{\partial x_j}$.

$$\frac{\partial}{\partial x_j} \circ a(x,D)f = \frac{\partial a}{\partial x_j}(x,D)f + a(x,D)\frac{\partial}{\partial x_j} f.$$

The second term on the right has the correct form. For the first term

$$\frac{\partial a}{\partial x_j}(x,\xi) = Q_x^1(\xi)a_{m-2}(x,\xi) + \sum_k a_{m-1}^{(k)}(x,\xi)\xi_k + a_m(x,\xi)$$

where $a_{m-2} \in S_\rho^{m-2}$, $a_{m-1}^{(k)} \in S_\rho^{m-1}$, $a_m \in S_\rho^m$. This is the correct form because

$$Q_x^1(\xi)a_{m-2} = \sum_j L_x^j(\xi)a_{m-2}\, L_x^j(\xi)$$

and $L_x^j(\xi)a_{m-2} \in S_\rho^m$ by Lemma 1.

We will prove Theorem 11 in cases of increasing generality.

Case 1. $m = 0$, k an integer. For $a \in S_\rho^0$ and $f \in L_k^p$ it suffices to check that $p(x,D) \circ a(x,D)f \in L^p$ for every polynomial $p(x,\xi)$ of degree $\leq k$. By Lemma 2, $p(x,D)a(x,D)f = \sum a_j(x,D)p_j(x,D)f \in L^p$, because of Theorem 10.

Case 2. $m = -2\ell$, ℓ, k integers.

For $a \in S_\rho^{-2\ell}$ we wish to show that $a(x,D): L_k^p \to L_{k+\ell}^p$. But the proof of Lemma 2 implies that $p(x,D) \circ a(x,D)$ has a symbol in S_ρ^0 if $\deg p(x,\xi) = \ell$. Hence we are reduced to Case 1.

Case 3. $m = -2\ell$, ℓ integer, k real ≥ 0

This is a consequence of Case 2 and complex interpolation. Calderón [8].) Complex interpolation is also used below.

Case 4. The general case m real ≤ 0, k real ≥ 0.

Suppose $a \in S_\rho^m$. Let ℓ be the integer such that $-2\ell-2<$

Denote $a_z(x,\xi) = a(x,\xi)\rho(x,\xi)^z$. Then $a_z \in S_\rho^{m+\mathrm{Re}\, z}$, (see t

in Appendix) and we interpolate in the strip $-2\ell - 2 - m \leq$ Re

$f = f_0 + f_1$ and $\|f_0\|_{B_0} \leq At^{\theta}$, $\|f_1\|_{B_1} \leq At^{-1+\theta}\}$. The smallest A above defines a norm for B_{θ}. A trivial consequence of the definition is that if $B_1 \subset B_0$ and $B_1' \subset B_0'$ and T is a bounded operator $B_0 \to B_0'$ and $B_1 \to B_1'$, then T is a bounded operator $B_{\theta} \to B_{\theta}'$.

For each integer $k \geq 0$, let $C^{(k)}$ denote k times continuously differentiable functions with norm

$$\|f\|_{C^{(k)}} = \sum_{|\alpha| \leq k} \left\| \frac{\partial^{\alpha} f}{\partial x^{\alpha}} \right\|_{L^{\infty}}.$$

Proposition 1. *$\Lambda_{\alpha} = [C^{(j)}, C^{(k)}]_{\theta}$, where $0 < \theta < 1$, $k > j \geq 0$ and $\alpha = \theta k + (1-\theta)j$.*

Proof. We will only look at the special case $\Lambda_{\theta} = [C^{(0)}, C^{(1)}]_{\theta}$. Suppose $f \in [C^{(0)}, C^{(1)}]_{\theta}$. It suffices to estimate $\|f(x-y) - f(x)\|_{L^{\infty}}$ for $|y| \leq 1$, because f is bounded. (From now on $|\cdot|$ denotes Euclidean norm unless otherwise noted.) For $t = |y|$, we can choose a decomposition $f = f_0 + f_1$. Then

$$\|f(x-y) - f(x)\|_{L^{\infty}} \leq \|f_0(x-y) - f_0(x)\|_{L^{\infty}} + \|f_1(x-y) - f_1(x)\|_{L^{\infty}}$$

$$\leq \|f_0(x-y)\|_{L^{\infty}} + \|f(x)\|_{L^{\infty}} + |y| \sum_j \left\| \frac{\partial f_1}{\partial x_j} \right\|_{L^{\infty}}$$

$$\leq 2At^{\theta} + A|y|t^{-1+\theta} \leq 3A|y|^{\theta}.$$

Conversely, suppose $f \in \Lambda_{\theta}$. Choose $\varphi \in C_0^{\infty}$ with $\int \varphi(x)\,dx = 1$. Denote $\varphi_{\epsilon}(x) = \epsilon^{-n}\varphi(x/\epsilon)$.

$$f = (f - f*\varphi_\varepsilon) + f*\varphi_\varepsilon = f_0 + f_1 .$$

Given $t \in (0,1]$, let $\varepsilon = t$. Then

$$|f_0(x)| = |f(x) - f*\varphi_\varepsilon(x)| = |\int (f(x-y) - f(x))\varphi_\varepsilon(y)\,dz|$$

$$\leq A\int |y|^\theta \varphi_\varepsilon(y)\,dy = A'\varepsilon^\theta = A't^\theta .$$

In order to estimate $\|f_1\|_{C^{(1)}}$ it is necessary to look at $\frac{\partial}{\partial x_j}(f*\varphi_\varepsilon(x))$. Here we use the fact that $\frac{\partial}{\partial x_j}\varphi_\varepsilon(x)$ has mean-value zero. The details are left to the reader.

Proposition 2. $[\Lambda_{\alpha_0}, \Lambda_{\alpha_1}]_\theta = \Lambda_\alpha$ where $\alpha = \theta_{\alpha_1} + (1-\theta)_{\alpha_0}$. (The proof is omitted but is very similar to that of Proposition 1.) For more details see Lions-Peetre [32], O'Neil [36], and Taibleson [43].

The simplest proof of Theorem 12 involves yet another closely related characterization of Λ_α.

Proposition 3. Choose α, $0<\alpha<\infty$ and an integer $k > \alpha$. Then $f \in \Lambda_\alpha$ if and only if $f = \sum_{j=0}^{\infty} f_j$ where $f_j \in C^{(k)}$ and $\|f_j\|_{L^\infty} \leq A\,2^{-j\alpha}$, and more generally,

$$\|\nabla^\ell f_j\|_{L^\infty} \leq A\,2^{-j(\alpha-\ell)} \quad \text{for } 0 \leq \ell \leq k.$$

Proof. This is essentially an extension of Proposition 1. For example, suppose $0<\alpha<1$ and $f = \sum_{j=0}^{\infty} f_j$ with $f_j \in C^{(1)}$, $\|f_j\|_{L^\infty} \leq A\,2^{-j\alpha}$, and $\|\nabla f_j\|_{L^\infty} \leq A\,2^{-j(\alpha-1)}$. Let $t = 2^{-k}$ and $f^{(0)} = \sum_{j=k+1}^{\infty} f_j$, $f^{(1)} = \sum_{j=0}^{k} f_j$.

Then $\|f^{(0)}\|_{C^{(0)}} \leq A t^{\alpha}$ and $\|f^{(1)}\|_{C^{(1)}} \leq A t^{-1+\alpha}$. Therefore $f \in \Lambda_{\alpha}$. The remainder of the proof is left to the reader. (See also the treatment of Γ_{α} below.)

The proof of Theorem 12 is a variant of the argument used in Rothschild-Stein [39] to deal with singular integrals related to nilpotent groups. It depends on

Main Lemma. *If $a(x, \xi) \in S^m_\rho$, $m < 0$, then $a(x, D)$ sends L^∞ into Λ_α for $\alpha = -m/2$.*

Proof. Recall the decomposition $a(x, \xi) = \sum a_j(x, \xi)$ where $a_j(x, \xi)$ is supported where $\rho(x, \xi) \simeq 2^j$.* Since $a_j(x, \xi)$ arises from a symbol of the class S^m_ρ (not S^0_ρ), we have the (analogous) property $2^{-mj} a_j(x, \xi) \in S^0_\rho$ uniformly in j. Let $a_j(x, D) f = \int K_j(x, x-z) f(z) dz$; then $2^{-mj} K_j(x, x-z)$ satisfies the same estimates as K_j in Lemma 2 (§6). Thus if we take λ to be a unit vector

$$|K_j(x, z)| \leq C_M 2^{jm} V_x(2^{-j})^{-1} \min (1, N_x(z) 2^j)^{-M},$$

$$|\nabla^{\ell}_x, K_z(x, z)| \leq C_M 2^{jm} V_x(2^{-j})^{-1} \min (1, N_x(z) 2^j)^{-M} 2^{2\ell j}.$$

Suppose $f \in L^\infty$; then

$$a(x, D) f = \sum \int K_j(x, x-z) f(z) dz = \sum g_j,$$

$$\|g_j\|_{L^\infty} \leq \|K_j\| \|f\|_{L^\infty} < A 2^{jm} = A 2^{-2j\alpha}.$$

Similarly,

$$\|\nabla^{\ell} g_j\|_{L^\infty} \leq A^{-2j(\alpha-\ell)}$$

*The definition of a_j is given after Lemma 1 in §6.

Let $f_j = \begin{cases} g_{j/2} & j/2 \text{ is an integer} \\ 0 & \text{otherwise} \end{cases}$

Then $a(x,D)f = \sum f_j \in \Lambda_\alpha$ for $\alpha = -m/2$.

The main lemma implies, in particular, that if $a \in S^m_\rho$, $m < 0$, then $a(m,D)$ maps $C^{(0)}$ to $\Lambda_{-m/2}$. Furthermore, using (1) we see that $a(x,D)$ maps $C^{(1)}$ to $\Lambda_{-m/2+1}$. By interpolation, (Propositions 1 and 2) $a(x,D)$ maps Λ_α to $\Lambda_{\alpha-m/2}$ for $0 < \alpha < 1$. A similar argument shows that $a(x,D)$ maps Λ_α to $\Lambda_{\alpha-m/2}$ when α is not an integer. A second interpolation (Proposition 2) fills in the integer values of α.

Remarks on the case $m=0$

1. If $a \in S^0_\rho$, it is not necessarily true that $a(x,D)$ maps Λ_α to Λ_α. It even fails for the heat equation, $Q(\xi) = \sum_{j=2}^{n} \xi_j^2$ (see Madych-Rivière [33]). One way to see this is to observe that a pseudo-differential operator acting in one variable alone (e.g., a partial Hilbert transform) does not preserve $\Lambda_\alpha(\mathbb{R}^2)$. A partial Hilbert transform is a limit of non-isotropic pseudo-differential operators on $L^2(\mathbb{R}^2)$.

2. On the other hand, suppose $a \in S^0_\rho$ and $a(x,D) = p(x,D) \circ b(x,D)$, where p is a polynomial of degree k and $b \in S^{-2k}_\rho$. Applying Theorem 12 to $b(x,D)$, we see that $a(x,D)$ maps Λ_α to Λ_α for $\alpha > 0$. For example, the Cauchy-Szegö operator has a symbol in S^0_ρ for appropriate ρ, but has this extra property. (See Chapter IV, also Phong and Stein [37].)

§9. Non-isotropic Sobolev and Lipschitz spaces

We shall define generalized Lipschitz spaces Γ_α $(0 < \alpha < \infty)$ with respect to the non-isotropic balls $B(x,\delta)$ as follows. Choose k so that $k > \alpha \geq k-1$. We say that $f \in \Gamma_\alpha$ if $f \in L^\infty$ and there exists A such that for all x_0 and $\delta > 0$ there exists a polynomial $P_{(x_0,\delta)}(x)$ of degree $\leq k-1$ such that

(2) $$|f(x) - P_{(x_0,\delta)}(x)| \leq A\,\delta^\alpha, \quad x \in B(x_0,\delta).$$

A norm for Γ_α is $\|f\|_{L^\infty}$ + least possible A above.

Remark. In what follows it is not hard to see that we could replace (2) by

$$\left(\frac{1}{V_{x_0}(\delta)} \int_{B(x_0,\delta)} |f(x) - P_{(x_0,\delta)}(x)|^p\,dx\right)^{1/p} \leq A\,\delta^\alpha$$

(This remark is motivated by the work of Campanato and Meyers, see also Krantz [30].)

Example. If $0 < \alpha < 1$, the triangle inequality implies that $f \in \Gamma_\alpha$ iff f is bounded and

$$|f(x) - f(y)| \leq C\,N_x(x-y)^\alpha.$$

Theorem 13. Suppose $\alpha > 0$, $\alpha - m > 0$, and $a \in S^m_\rho$. Then $a(x,D)$ is bounded $\Gamma_\alpha \to \Gamma_{\alpha-m}$.

Main Lemma. $f \in \Gamma_\alpha$ if and only if $f = \sum_{j=0}^{\infty} f_j$ where $f_j \in C^\infty$.

(3) $$\|f_j\|_{L^\infty} \leq A\,2^{-j\alpha} \quad \text{and more generally}$$

(4) $$\left|(\lambda_1, \tfrac{\partial}{\partial x}) \cdots (\lambda_\ell, \tfrac{\partial}{\partial x}) f_j(x)\right| \leq A\,2^{-j\alpha} \prod_{i=1}^{\ell} \theta(N_x(\lambda_i)2^j) \quad \text{for } 0 \leq \ell \leq k.$$

$(k > \alpha)$.

Proof. Suppose $f \in \Gamma_\alpha$. We will partition f using

$$a_j(x,\xi) = \psi(2^{-j}\rho(x,\xi)) - \psi(2^{-j+1}\rho(x,\xi)), \quad a_0(x,\xi) = \psi(\rho(x,\xi)),$$

$$f = \sum a_j(x,D)f = \sum_{j=0}^{\infty} f_j.$$

Now $f_j(x) = \int K_j(x, x-y) f(y)dy$ and K_j is orthogonal to any polynomial in y for $j \neq 0$. Thus the natural thing to do is replace $f(y)$ by $f(y) - P_{(x,2^{-j})}(y)$. First we prove some lemmas.

Lemma 1. *If* $f \in \Gamma_\alpha$ *and* $t \geq 1$, *then* $|f(x) - P_{(x_0,\delta)}(x)| \leq A t^a \delta^\alpha$ *for* $x \in B(x_0, t\delta)$ *and some fixed* a.

Lemma 2. *If* P *is a polynomial of degree* $k-1$, $k \geq 1$, *then*

$$\sup_{x \in B(x_0, t\delta)} |P(x)| \leq C t^{2k-2} \sup_{x \in B(x_0,\delta)} |P(x)|, \qquad t \geq 1.$$

It suffices to prove Lemma 2 assuming that $B(x_0, \delta)$ is an ellipse, since it is trapped between two similar ellipses. A linear transformation preserves the class of polynomials (of degree $k-1$), so we can pass from ellipses to the ordinary unit ball. Here Lemma 2 is obvious, if we use the fact that $B(x_0, t\delta)$ is essentially contained in the dilate of $B(x_0,\delta)$ by the factor t^2 (see the proof of Proposition 5 in Chapter I).

As for Lemma 1,

$$|f(x) - P_{(x_0,\delta)}(x)| \leq A\delta^\alpha \qquad x \in B(x_0, \delta)$$

$$|f(x) - P_{(x_0,t\delta)}(x)| \leq A t^\alpha \delta^\alpha \qquad x \in B(x_0, t\delta)$$

Thus $|P_{(x_0,t\delta)}(x) - P_{(x_0,\delta)}(x)| \leq 2At^\alpha\delta^\alpha, \quad x \in B(x_0, \delta)$.

Now apply Lemma 2 to $P_{(x_0, t\delta)}(x) - P_{(x_0, \delta)}(x)$ to obtain Lemma 1 with $a = \alpha + 2k - 2$.

Return to the main lemma.

$$(\lambda_1, \frac{\partial}{\partial x}) \ldots (\lambda_\ell, \frac{\partial}{\partial x}) f_j(x) = \int (\lambda_1, \frac{\partial}{\partial x}) \ldots (\lambda_\ell, \frac{\partial}{\partial x}) K_j(x, x-y) f(y)\, dy$$

$$= \int (\lambda_1, \frac{\partial}{\partial x}) \ldots (\lambda_\ell, \frac{\partial}{\partial x}) K_j(x, x-y)(f(y) - P_{(x, 2^{-j})}(y))\, dy$$

Since $a_j(x, \xi) \in S^0_\rho$ uniformly in j, then by the analogue of Lemma 2 in §6,

$$|(\lambda_1, \frac{\partial}{\partial x}) \ldots (\lambda_\ell, \frac{\partial}{\partial x}) K_j(x, x-y)| \lesssim C_M V_x(2^{-j})^{-1} \min(1, 2^j N_x(x-y))^{-M} \prod_{i=1}^{\ell} \theta(N_x(\lambda_i) 2^j).$$

Combined with Lemma 1 for $\delta = 2^{-j}$ and $t = 2^m$:

$$|(\lambda_1, \frac{\partial}{\partial x}) \ldots (\lambda_\ell, \frac{\partial}{\partial x}) f_j(x)| \leqq$$

$$\lesssim \prod_{i=1}^{\ell} \theta(N_x(\lambda_i) 2^j) \Big\{ \int_{B(x, 2^{-j})} V_x(2^{-j})^{-1} (2^{-j})^{\alpha} dy + \sum_{m=1}^{\infty} \int_{B(x, 2^{-j+m}) \setminus B(x, 2^{-j+m-1})} V_x(2^{-j})^{-1} 2^{ma} 2^{-j\alpha} 2^{-mM} dy \Big\}$$

$$\leqq \Big[\prod_{i=1}^{\ell} \theta(N_x(\lambda_i) 2^j) \Big] 2^{-j\alpha}$$

for sufficiently large M.

Conversely, if $f = \sum_{j=0}^{\infty} f_j$, with f_j satisfying (3) and (4). Denote

$F_\delta(x) = \sum_{2^{-j} \geq \delta} f_j(x)$. The sum is finite, so $F_\delta \in C^\infty$.

$$\|f - F_\delta\|_{L^\infty} \leqq \sum_{2^{-j} < \delta} 2^{-j\alpha} \leq C\delta^\alpha.$$

Therefore, it suffices to find a polynomial approximation to F_δ. Let

$P_{(x_0,\delta)}(y)$ be the Taylor polynomial of degree $k-1$ at x_0 of F_δ. Without loss of generality $x_0 = 0$.

$$\Big|F_\delta(y) - \sum_{|\beta| \leq k-1} a_\beta y^p\Big| = \Big|\frac{1}{(k-1)!}\int_0^1 (1-t)^{k-1} (y, \frac{\partial}{\partial x})^k F_\delta(ty)\,dt\Big|$$

$$\leq A \sum_{2^{-j} \geq \delta} 2^{-j\alpha} (N_0(y)\, 2^j)^k$$

$$\leq A \sum_{j=0}^{\log_2 \delta^{-1}} \delta^k 2^{j(k-\alpha)} \leq A\,\delta^\alpha ,$$

because $k > \alpha$ and $N_0(y) \leq \delta$.

The proof of Theorem 13 is now just a repetition of the main lemma. If $a \in S^m_\rho$ and $f \in \Gamma_\alpha$, then $a(x,D)f = \Sigma\, a_j(x,D)f = \Sigma\, g_j$, where $2^{-jm} a_j(x,\xi) \in S^0_\rho$ uniformly in j. It is easy to check that g_j satisfies the conditions of the main lemma in order that $a(x,D)f = \Sigma\, g_j \in \Gamma_{\alpha - m}$.

Remark 1. The proof shows that in the classical case, $Q(x,\xi) = \|\xi\|^2$, Γ_α is the same as Λ_α.

Remark 2. We have the inclusion relations

$$\Lambda_\alpha \subset \Gamma_\alpha \subset \Lambda_{\alpha/2}.$$

If $B_0(x,\delta)$ is the standard Euclidean ball of radius δ then it is easy to see that $B_0(x,\delta^2) \subsetsim B(x,\delta) \subsetsim B_0(x,\delta)$, $0 < \delta \leq 1$, (see Proposition 2). So the inclusion $\Lambda_\alpha \subset \Gamma_\alpha \subset \Lambda_{\alpha/2}$ follows easily, if we appeal to Remark 1.

Remark 3. Variants of the Γ_α spaces have been studied in other contexts (see Stein [42], Folland-Stein [16], Folland [14]). It would be interesting to give the precise relation between those spaces, and the type we have considered above. In this connection see also Krantz [30].

We now pass to non-isotropic Sobolev spaces. We shall need an extra assumption on ρ, which will also play a crucial role in Chapter IV.

A_1 { There exists $P(x,\xi) \in S^1_\rho$, $Q(x,\xi) \in S^{-1}_\rho$ such that $Q(x,D) \circ P(x,D) - \mathrm{Id.}$ has a symbol in S^{-1}_ρ and $P(x,\xi) \in S^1_{1,0}$, (the space of classical first order symbols).

More generally, it will suffice to have a collection of symbols $P_j \in S^1_\rho \cap S^1_{1,0}$, $Q_j \in S^{-1}_\rho$ such that $\sum P_j \circ Q_j - \mathrm{Id.}$ has a symbol in S^{-1}_ρ. In applications, our symbol classes will arise from parametrices of explicit operators (e.g., parametrices for $\Box_b$, when there exists operators with symbols in the class S_ρ). This will make it possible to satisfy A_1. Notice that in order to satisfy A_1 we only need

$$Q \circ P = I - R \quad \text{for} \quad R \in S^{-\epsilon}_\rho \quad \text{for some} \quad \epsilon > 0.$$

For then $(1 + R + \ldots + R^{M-1}) Q \circ P = I - R^M$ and R^M has a symbol in S^1_ρ for sufficiently large M (see Theorem 17).

Sometimes we shall only be able to obtain the weaker condition A_2, identical to A_1 except that S^1_ρ, S^{-1}_ρ, $S^1_{1,0}$ are replaced by S^2_ρ, S^{-2}_ρ, $S^2_{1,0}$.

Definition. $f \in \mathrm{Sob}^p_k$ iff $a(x,D) f \in L^p$ for all $a \in S^k_\rho$. Under assumption A_1, we will only consider Sob^p_k for integers $k \geq 0$. Under assumption A_2, we consider only even integers $k \geq 0$.*

<u>Theorem 14.</u> $f \in Sob_k^p \iff f \in L^p$ <u>and</u> $P^k f \in L^p$ <u>where</u> P <u>is defined in</u> A_1. <u>In the case of</u> A_2 <u>the condition is</u> $P^{k/2} f \in L^p$, k <u>even</u>. (In general, we replace P^k by every monomial of degree k in the operators P_j.)

<u>Proof.</u> We will assume throughout that if $p_1 \in S_\rho^{m_1}$, $p_2 \in S_\rho^{m_2}$, then $p_1(x,D) \circ p_2(x,D)$ has a symbol in $S_\rho^{m_1+m_2}$. This will be proved in the next section (Theorem 17).

It follows from A_1 that there are operators Q_k and R_k with symbols in S_ρ^{-k} such that

$$Q_k P^k = I - R_k$$

(e.g., let $A = I + R + \ldots + R^M$ and let $Q_k = (AQ)^k$, where Q and R are defined in A_1.) Therefore,

$$a(x,D)f = [a(x,D)Q_k]P^k f + [a(x,D)R_k]f \in L^p,$$

since $a(x,D)Q_k$ and $a(x,D)R_k$ have symbols in S_ρ^0.

<u>Theorem 15.</u> <u>If</u> $a(x,\xi) \in S_\rho^m$, <u>then</u> $a(x,D)$ <u>is bounded from</u> Sob_k^p <u>to</u> Sob_{k-m}^p. This is an immediate consequence of the definition of Sob_k^p and Theorem 17.

<u>Theorem 16.</u> $L_k^p \subset Sob_k^p \subset L_{k/2}^p$ <u>(locally)</u>

<u>Proof.</u> We only consider the case $k=1$, the other cases are similar.

*The spaces Sob_k^p are the analogues of the spaces S_k^p considered in [16] and [39].

(a) $L_1^p \subset Sob_1^p$. Suppose $f \in L_1^p$, then $f \in L^p$ and $Pf \in L^p$, because P has a symbol in $S^1_{1,0}$. Hence by Theorem 14, $f \in Sob_1^p$.

(b) $Sob_1^p \subset L^p_{1/2}$. Suppose that $f \in Sob_1^p$. Denote by T_z the operator whose symbol is $\rho(x,\xi)^z$. Recall that $\rho(x,\xi)^z \in S_\rho^{\mathrm{Re}\, z}$. (See Appendix.)

Denote $F_z = T_z f$. We will interpolate in the strip $-1 \leq \mathrm{Re}\, z \leq 1$. For $z = 1+iy$,

$$F_z = T_z(QP-R)f = (T_zQ)Pf - (T_zR)f \in L^p,$$

because T_zQ and T_zR have symbols in S_ρ^0. For $x = -1+iy$, T_zQ and T_zR have symbols in S_ρ^{-2}. Thus $\frac{\partial}{\partial x_j} \circ T_zQ$ and $\frac{\partial}{\partial x_j} \circ T_zR$ has symbols in S_ρ^0. (See Lemma 1 to Theorem 11.) Furthermore the bounds on $\|F_z\|_{L^p}$ for $z = 1+iy$ and $\|F_z\|_{L_1^p}$ for $z = -1+iy$ have polynomial growth in y. By interpolation, $f = T_0 f \in L^p_{1/2}$.

__Remark.__ The reason why interpolation is necessary in the proof above is that although with respect to its first m derivatives $\|\xi\|^m$ behaves like a symbol in S_ρ^{2m}, $\|\xi\|^m \notin S_\rho^{2m}$ unless m is an (even) integer.*

§10. The composition of operators

__Theorem 17.__ __Suppose__ $a \in S_\rho^{m_1}$, $b \in S_\rho^{m_2}$. __Then__ $a(x,D) \circ b(x,D) = c(x,D)$ __where__ $c \in S^{m_1+m_2}$.

For the standard symbols, and more generally those in the class

*This remark is one of the ideas behind the definition of the extended symbol class ES_ρ used in §15 below.

$S^m_{\rho,\delta}$ with $\rho > \delta$, then there is an asymptotic formula for c. This is no longer so when $\rho = \delta$ (in particular when $\rho = \delta = 1/2$). In that case the weaker result, the analogue of Theorem 17, was proved by Beals and Fefferman [3] and Boutet de Monvel [5]. It is their method we use here.

It is easy to obtain formally

$$c(x,\xi) = \iint a(x,\eta)\, b(y,\xi)\, e^{2\pi i(y-x)\cdot(\eta-\xi)}\, dy\, d\eta . \tag{5}$$

The integral converges absolutely if $a(x,\eta)$ and $b(y,\xi)$ are supported in annuli in η and ξ. For general symbols formula (5) has meaning as a so-called oscillatory integral, whose value is computed by a formal integration by parts (see ${}^tL$ below).

We will prove Theorem 17 in the case $\varphi(x) \equiv 1$ only.

<u>Principal Lemma.</u> <u>Suppose that</u>

$$c(x,\xi) = \iint c(x,y,\eta,\xi)\, e^{2\pi i(y-x)\cdot(\eta-\xi)}\, dy\, d\eta$$

<u>and there exist functions</u> $C(x,\xi)$ <u>and</u> $C(x,y,\eta,\xi)$ <u>such that</u>

(i) $$|\partial^\alpha_\eta \partial^\beta_y c(x,y,\eta,\xi)| \le \text{const}\, C(x,y,\eta,\xi)\, |\eta|^{-|\alpha|/2}\, |\xi|^{|\beta|/2},$$ *

(ii) $$C(x,y,\eta,\xi) \le \text{const}\,(1+|\xi|+|\eta|)^{M_1}\, C(x,\xi),$$

(iii) <u>For some</u> $\epsilon > 0$

$$|\xi-\eta| \le \epsilon(\rho(x,y,\eta) + \rho(x,\xi))^{\dagger}$$

---

* $|\cdot|$ denotes Euclidean norm from now on.

† In our first application, $\rho(x,y,\eta) = (\rho(x,\eta)\,\rho(x,\xi))^{1/2}$, but for the proof of the lemma $\rho(x,y,\eta)$ can be any non-negative function.

<u>and</u> $|x-y| \le \epsilon(\frac{1}{|\eta|}\rho(x,y,\eta) + \frac{1}{|\xi|}\rho(x,\xi))$

<u>implies</u> $C(x,y,\eta,\xi) \le \text{const } C(x,\xi)$.

(iv) <u>If</u> $|\eta| \simeq |\xi|$, <u>then</u>

$$C(x,y,\eta,\xi) \le \text{const}\left(\frac{\rho(x,y,\eta)^2}{|\eta|} + \frac{\rho(x,\xi)^2}{|\xi|}\right)^{M_2} C(x,\xi).$$

<u>Then we can conclude that</u> $|c(x,\xi)| \lesssim C(x,\xi)$.

Our hypotheses can be described roughly as follows. In the region where ξ and η are far apart we need only a weak estimate (ii). Where $|\xi| \simeq |\eta|$ we need a stronger estimate (iv), but where ξ is close to η and x is close to y we need the strongest estimate (iii). This should be clear from the proof.

Denote $D = I - \alpha\Delta_y - \beta\Delta_\eta$, where

$$\alpha = \frac{1}{(1+|\xi|^2)^{1/2}} + \frac{1}{(1+|\eta|^2)^{1/2}} \text{ and } \beta = (1+|\xi|^2)^{1/2} + (1+|\eta|^2)^{1/2}$$

$$e = e(x,y,\eta,\xi) = e^{2\pi i(y-x)(\eta-\xi)}.$$ Finally

$$Q = (1 + 4\pi^2\alpha|\xi-\eta|^2 + 4\pi^2\beta|x-y|^2)$$

Notice that $De = Qe$. Thus, denoting $L = Q^{-1}D$, $Le = e$. Therefore

(6) $L^N e = e$.

This basic identity allows us to integrate by parts.

$$c(x,\xi) = \int\int [({}^tL)^N c(x,y,\eta,\xi)]\, e\, dy\, d\eta$$

Hence,

(7) $$|c(x,\xi)| \leq \iint |({}^tL)^N c(x,\xi,\eta,\xi)|\, dy\, d\eta$$

As a consequence of (i), one can show

(8) $$|({}^tL)^N c(x,y,\eta,\xi)| \lesssim Q^{-N}\left(1+\frac{|\xi|}{1+|\eta|}+\frac{|\eta|}{1+|\xi|}\right)^N C(x,y,\eta,\xi)$$

In fact, ${}^tL = {}^tDQ^{-1}$, so that $({}^tL)^N c(x,y,\eta,\xi) = ({}^tDQ^{-1})^N c(x,y,\eta,\xi)$ involves large numbers of terms. For example, when all differentiation falls on $c(x,y,\eta,\xi)$ we get terms like

$$|Q^{-N}\alpha^{N_1}\beta^{N_2}\Delta_y^{N_1}\Delta_\eta^{N_2} c(x,y,\eta,\xi)|$$

$$\lesssim Q^{-N}\alpha^{N_1}\beta^{N_2}|\eta|^{-N_2}|\xi|^{N_1} C(x,y,\eta,\xi) \lesssim Q^{-N}\left(1+\frac{|\xi|}{1+|\eta|}\right)^N C(x,y,\eta,\xi)$$

where $N_1+N_2=N$. The rest of the verification of (8) is left to the reader.

We will break up the analysis of (7) into three regions.

Region I: $|\xi-\eta| \geq \frac{1}{2}(1+|\xi|+|\eta|)$.

In this case

$$\begin{aligned} Q &\geq \left(\left(\frac{1}{1+|\xi|}\right)+\frac{1}{1+|\eta|}\right)(1+|\xi|+|\eta|)^2 \\ &\geq \left(1+\frac{|\xi|}{1+|\eta|}+\frac{|\eta|}{1+|\xi|}\right)(1+|\xi|+|\eta|) \\ &\geq \left(1+\frac{|\xi|}{1+|\eta|}+\frac{|\eta|}{1+|\xi|}\right)^{3/2}(1+|\xi|+|\eta|)^{1/2} \end{aligned}$$

because $ab \geq a^{3/2}b^{1/2}$ when $a < b$. Hence,

$$Q \geq \left(1+\frac{|\xi|}{1+|\eta|}+\frac{|\eta|}{1+|\xi|}\right)(1+|\xi|+|\eta|)^{1/3}\, Q^{1/3}.$$

And therefore,

$$Q^{-N}\left(1+\frac{|\xi|}{1+|\eta|}+\frac{|\eta|}{1+|\xi|}\right)^{N} \lesssim Q^{-N/3}\,(1+|\xi|+|\eta|)^{-N/3}.$$

For sufficiently large N, (8) and (ii) now show that

$$\iint_{\text{Region I}} |({}^{t}L)^{N} c(x,y,\eta,\xi)|\,dy\,d\eta \qquad \text{is dominated by}$$

$$\left(\iint Q^{-N/3}\,dy\,d\eta\right) C(x,\xi).$$

Thus we need only prove that

(9) $$\iint Q^{-N/3}\,dy\,d\eta \leqq \text{const. for sufficiently large } N.$$

In fact, let $a = 1/(1+|\xi|^2)^{1/2}$. Then $\alpha > a$ and $\beta > a^{-1}$ implies

$$Q \geq 1 + a|\xi-\eta|^2 + a^{-1}|x-y|^2.$$

If $N/3 > n$, then

$$\iint Q^{-N/3}\,dy\,d\eta \leq \iint \frac{dy\,d\eta}{(1+a|\xi-\eta|^2+a^{-1}|x-y|^2)^{N/3}} = \iint \frac{dy\,d\eta}{(1+|\xi-\eta|^2+|x-y|^2)^{N/3}} \leqq \text{const.}$$

(The change of variables η to $a^{-1/2}\eta$ and y to $a^{1/2}y$ has Jacobian determinant $\equiv 1$.)

In the complement of Region I, $|\xi-\eta| < \frac{1}{2}(1+|\xi|+|\eta|)$, we have $|\xi| \simeq |\eta|$. For this reason, (8) reduces to

(8′) $$|({}^{t}L)^{N} c(x,\xi,\eta,\xi)| \lesssim Q^{-N} C(x,\xi,\eta,\xi)$$

<u>Region II</u>: The complement of Region I intersect the region where (iii) holds

$$\iint_{\text{II}} |({}^{t}L)^{N} c(x,\xi,\eta,\xi|\,dy\,d\eta \leq \text{const.} \iint_{\text{II}} Q^{-N} C(x,\xi,\eta,\eta)\,dy\,d\eta \leqq$$

$$\lesssim \text{const.} \left(\iint Q^{-N}\, dz\, d\eta\right) C(x,\xi) \lesssim \text{const.}\; C(x,\xi)$$

by (9) and (iii).

<u>Region III:</u> $|\xi| \simeq |\eta|$, but one of the conditions of (iii) fails.

Suppose that $|\xi-\eta| > \text{const.}\ (\rho(x,y,\eta) + \rho(x,\xi))$.

Then $Q \geq \alpha|\xi-\eta|^2 \geq c(1/(1+|\xi|) + 1/(1+|\eta|))(\rho(x,y,\eta)^2 + \rho(x,\xi)^2)$

$$\gtrsim \left(\frac{\rho(x,y,\eta)^2}{|\eta|} + \frac{\rho(x,\xi)^2}{|\xi|}\right).$$

For large N, Q^{-N} dominates $\left(\frac{\rho(x,y,\eta)^2}{|\eta|} + \frac{\rho(x,\xi)^2}{|\xi|}\right)^{M_2}$

with enough extra powers to space as to be integrable as in Regions I and II. When $|x-y| > \text{const.} \left(\frac{\rho(x,y,\eta)}{|\eta|} + \frac{\rho(x,\xi)}{|\xi|}\right)$, the proof is similar.

Let us return to $c(x,\xi) = a(x,\xi) \circ b(x,\xi)$. To obtain $|c(x,\xi)| \leq \text{const.}\ \rho(x,\xi)^{m_1+m_2}$, we will apply the principal lemma with $\rho(x,y,\eta) = (\rho(x,\eta)\,\rho(y,\eta))^{1/2}$,

$$c(x,y,\eta,\xi) = a(x,\eta)\, b(y,\xi)$$

$$C(x,\xi) = \rho(x,\xi)^{m_1+m_2}$$

(10)
$$C(x,y,\eta,\xi) = \rho(x,\eta)^{m_1}\, \rho(y,\xi)^{m_2}$$

Since $S^m_\rho \subset S^m_{1/2,1/2}$ (assuming $\varphi(x) \equiv 1$), the estimates in (i) are easy to obtain.

If $C_1(x,y,\eta,\xi)$ and $C_1(x,\xi)$ satisfy (ii) - (iv) and $C_2(x,y,\eta,\xi)$ and

$C_2(x,\xi)$ satisfy (ii) - (iv), then so do $C_1(x,y,\eta,\xi)^{a_1}\,C_2(x,y,\eta,\xi)^{a_2}$ and

$$C_1(x,\xi)^{a_1}\,C_2(x,\xi)^{a_2} \quad \text{for } a_1, a_2 \geq 0. \text{ Therefore,}$$

it suffices to check (ii) - (iv) for (10) in the cases $m_1 = 0$, $m_2 = \pm 1$ and $m_1 = \pm 1$, $m_2 = 0$. We will carry out the case $m_1 = 1$, $m_2 = 0$, i.e.,

$$C(x,\xi) = \rho(x,\xi)$$

$$C(x,y,\eta,\xi) = \rho(x,\eta)$$

The other cases are similar. The proof of (ii) is trivial.

(iv) When $|\xi| \simeq |\eta|$, we wish to show that

$$\rho(x,\eta) \lesssim \left(\frac{\rho(x,\eta)\,\rho(y,\eta)}{|\eta|} + \frac{\rho(x,\xi)^2}{|\xi|}\right)\rho(x,\xi).$$

This follows easily from $\rho(y,\eta)\,\rho(x,\xi) \gtrsim |\eta|$.

(iii) We must prove that $\rho(x,\eta) \lesssim \rho(x,\xi)$ under the hypotheses

(a) $|\xi-\eta| <$ small const. $(\rho(x,\eta)^{1/2}\,\rho(y,\eta)^{1/2} + \rho(x,\xi))$

(b) $|x-y| <$ small const. $\left(\frac{\rho(x,\eta)^{1/2}\,\rho(y,\eta)^{1/2}}{|\eta|} + \frac{\rho(x,\xi)}{|\xi|}\right)$.

The triangle inequality and (a) yield

$$\begin{aligned}\rho(x,\eta) &\lesssim \rho(x,\xi) + \text{const. } |\xi-\eta| \\ &\lesssim \rho(x,\xi) + \epsilon\,\rho(x,\eta)^{1/2}\,\rho(y,\eta)^{1/2}\end{aligned}$$

We will verify presently that

(11) $$\rho(y,\eta) \lesssim \rho(x,\eta) + \rho(x,\xi).$$

Using (11) and ε sufficiently small, we see that $\rho(x,\eta) \lesssim \rho(x,\xi)$.

An easy consequence of the formula $z\rho^3 \nabla_x \rho(x,\eta) = 2Q(x,\eta)\nabla_x Q(x,\eta)$ and $\rho(x,\eta)^2 \gtrsim |\eta|$ is the estimate $\sup_{x'} |\nabla_x \rho(x',\eta)| \leq$ const. $|\eta|$, where x' is between x and y. Hence

$$\rho(y,\eta) \lesssim \rho(x,\eta) + |x-y|\ |\eta| .$$

Combined with b) this yields (recall that $|\eta| \simeq |\xi|$)

$$\rho(y,\eta) \lesssim \rho(x,\eta) + \varepsilon(\rho(x,\eta)^{1/2}\rho(x,\eta)^{1/2} + \rho(x,\xi))$$

For ε sufficiently small, this implies (11).

Next, look at derivatives in ξ.

$(\eta, \frac{\partial}{\partial\xi})\, c(x,\xi) = ((\eta, \frac{\partial}{\partial\xi})a) \circ b + a \circ ((\eta, \frac{\partial}{\partial\xi})b)$. Therefore, a similar argument using appropriate rates of decrease for $(\eta, \frac{\partial}{\partial\xi})b$ gives estimates for $(\eta, \frac{\partial}{\partial\xi})\, c(x,\xi)$, and similarly for higher derivatives in ξ.

For derivatives in x there is a similar formula

$\frac{\partial}{\partial x_j}(a \circ b) = (\frac{\partial}{\partial x_j} \circ a) \circ b) + a \circ (\frac{\partial}{\partial x_j} \circ b)$. The symbol of $\frac{\partial}{\partial x_j} \circ a$ is (up to a factor of $-2\pi i$) $\xi_j \circ a_j(x,\xi) = a(x,\xi)\xi_j + \frac{\partial}{\partial x_j} a(x,\xi)$. Moreover,

$\frac{\partial a}{\partial x_j}(x,\xi) = \sum_{k=1}^{n} a_k(x,\xi)\xi_k + a_o(x,\xi)$ where $a_k \in S_\rho^{m_1 - 1}$ and $a_o \in S_\rho^{m_1}$.

Therefore, $(\frac{\partial a}{\partial x_j}) \circ b = \sum a_k \circ \xi_k \circ b + a_o \circ b$

$$= \sum a_k \circ (b\xi_k + \frac{\partial b}{\partial x_k}) + a_o \circ b$$

$$= \sum a_k \circ b\xi_k + \sum_{k,\ell} (a_k \circ b_\ell)\xi_\ell + \sum_k a_k \circ b_o + a_o \circ b$$

where $b_\ell \in S_\rho^{m_2-1}$. From our previous estimates on ξ derivatives we know that

$$a_k \circ b \in \hat{S}_\rho^{m_1+m_2-1}, \quad a_k \circ b_\ell \in \hat{S}_\rho^{m_1+m_2-2} \subset \hat{S}_\rho^{m_1+m_2-1}, \quad \text{and}$$

$\sum a_k \circ b_o + a_o \circ b \in \hat{S}_\rho^{m_1+m_2}$. It is clear that we can continue getting estimates taking higher x derivatives of $a_k \circ b$, $a_k \circ b_\ell$ etc. qed.

The adjoint $a(x,D)^*$ of a pseudo-differential operator $a(x,D)$ is defined by

$$\int a(x,D) f \bar{g}\, dx = \int f\, \overline{a(x,D)^* g}\, dx, \quad \text{for } f, g \in \mathcal{S}.$$

For example, if $a(x,\xi) = a(x)\, b(\xi)$, its formal adjoint is the composition $\overline{b(\xi)} \circ a(x)$. As this simple example shows, we can't expect the symbol of $a(x,D)^*$ to have compact support in x. We will define the symbol of the formal adjoint $a^{\#}(x,\xi)$ shortly. Since it does not have compact support, the best we can say is

Proposition 4. If $a \in S_\rho^m$, then $\psi(x)\, a^{\#}(x,\xi) \in S_\rho^m$ for any $\psi \in C_0^\infty$.

To derive a formula for $a^{\#}$, recall that

$$a(x,D) f(x) = \iint a(x,\xi)\, e^{-2\pi i (x-y)\cdot\xi} f(y)\, dz\, d\xi.$$

In light of our experience with $a(x)\, b(\xi)$ we might try to switch the order of multiplication in x and ξ.

$$(12) \qquad a(x,D)^* f(x) = \iint \overline{a(y,\xi)}\, e^{-2\pi i (x-y)\cdot\xi} f(y)\, dy\, d\xi$$

This leads one to consider a compound symbol that allows multiplication in x then in ξ then in x again. <u>A compound symbol</u> $a(x, \xi, y)$ acts on f by

$$a(x, D, y) f(x) = \iint a(x, \xi, y) e^{-2\pi i(x-y)\cdot\xi} f(y)\, dy\, d\xi$$

One should be aware that an operator has many "compound symbols," although it has a unique symbol.

The formal adjoint of $a(x, \xi, y)$ is clearly $\overline{a(y, \xi, x)}$. In particular, the formal adjoint of $a(x, D)$ has compound symbol $a(x, \xi, y) = \overline{a(y, \xi)}$. (This is just a fancy way of restating (12).) The point is now to find the ordinary symbol of $a(x, D, y)$. The relation is

$$a(x, \xi) = \iint a(x, \eta, y)\, e^{2\pi i(y-x)\cdot(\eta-\xi)}\, dy\, d\eta .$$

This can be proved formally, and makes sense as an oscillatory integral like the one that appears in the product formula. Thus

$$(13) \qquad a^{\#}(x, \xi) = \iint \overline{a(y, \eta)}\, e^{2\pi i(y-x)\cdot(\eta-\xi)}\, dy\, d\eta$$

In order to show that $\psi(x)\, a^{\#}(x, \xi) \in S^m_\rho$, we apply our principal lemma. For example, to estimate the growth of $\psi(x)\, a^{\#}(x, \xi)$, use

$$C(x, y, \eta, \xi) = \rho(y, \eta)^m, \quad C(x, \xi) = \rho(x, \xi)^m$$

§11. <u>A Fourier integral operator; change of variables</u>

Let us consider a rudimentary Fourier integral operator

$$I(f) = \iint e^{2\pi i\, \Phi(x-y)\cdot\xi}\, a(x, \xi)\, \eta(y)\, f(y)\, dy\, d\xi ,$$

where

(14) $a \in S^m_{\rho_0}$

(15) $\Phi = \mathbb{R}^n \times \mathbb{R}^n \longrightarrow \mathbb{R}^n$ is smooth; $\Phi(x,x) = 0$ and

$\nabla_y \Phi(x,y)\big|_{y=x} = M(x)$ is a non-singular $n \times n$ matrix.

(16) $\Phi(x,y)$ vanishes only when $x = y$. $\eta(y)$ is a C_0^∞ cut-off function.

If $\Phi(x,y) = -(x-y)$, then essentially $I(f) = a(x,D)f$.

Even in general, $I(f)$ is a pseudo-differential operator.

__Theorem 18.__ $I(f) = b(x,D)f$ __with__ $b \in S^m_{\tilde\rho}$ __where__

$$\tilde\rho(x,\xi) = \rho_0(x, \tilde M(x)\xi), \quad (\text{here } \tilde M = {}^t M^{-1})$$

__Observation.__ For x near y we can write

$$\Phi(x,y) = M(x,y)(y-x) \quad \text{where}$$

$M(x,y)$ is a smooth non-singular matrix-valued function and $M(x,x) = M(x)$.

To prove this, let $F(x,x-y) = \Phi(x,y) + M(x)(x-y)$. F is smooth and $F(x,0) = 0$, $\nabla_z F(x,z)\big|_{z=0} = 0$. Let $g(t) = F(x,tz)$, and use $g(1) = g(1) - g(0) - g'(0) = \int_0^1 (1-t)\, g''(t)\,dt$ to deduce that $F(x,z) = \sum_{i,j} z_i z_j F_{ij}(x,z)$ for smooth functions F_{ij}. Thus

(17) $$F(x,x-y) = \sum_{i,j} F_{i,j}(x,y)(x_i - y_i)(x_j - y_j)$$

Define $M_j(x,y) = M_j(x) - \sum_i F_{i,j}(x,y)(x_i - y_i)$, where M_j denotes the jth column of M. The equality $\Phi(x,y) = M(x,y)(y-x)$ follows from (17).

Choose a cut-off function $\chi \in C_0^\infty(\mathbb{R}^n)$ with $\chi(z) \equiv 1$ for small z.

$I(f) = I_1(f) + I_2(f)$, where

$$I_2(f) = \iint e^{2\pi i \Phi(x,y)\cdot\xi}(1 - \chi(x-y))\, a(x,\xi)\, \eta(y)\, f(y)\, dy\, d\xi$$

The kernel of the operator I_2

$$K(x,y) = \eta(y) \int e^{2\pi i \Phi(x,y)\cdot\xi}(1 - \chi(x-y))\, a(x,\xi)\, d\xi$$

can be defined by integration by parts (using (16)). Moreover, the same technique can be used to show that $K(x,y)$ is smooth (recall $a(x,\xi)$ has compact x-support). Thus $I_2(f)$ is a trivial error term.

In the study of $I_1(f)$ we assume that the support of $\chi(x-y)$ is sufficiently small that $\Phi(x,y) = M(x,y)(y-x)$ on the support of χ, and $\eta(y) = 1$.

$$I_1(f)(x) = \iint e^{-2\pi i M(x,y)(x-y)\cdot\xi}\, \chi(x-y)\, a(x,\xi)\, f(y)\, dy\, d\xi$$

$$= \iint e^{-2\pi i (x-y)\cdot {}^t M(x,y)\xi}\, \chi(x-y)\, a(x,\xi)\, f(y)\, dy\, d\xi$$

$$= \iint e^{-2\pi i (x-y)\cdot\xi}\, \chi(x-y)\, a(x, \widetilde{M}(x,y)\xi)\, f(y)\, |\det \widetilde{M}(x,y)|\, dy\, d\xi$$

Thus we have $I_1(f)(x) = a(x,D,y)\, f(x)$ where

$$a(x,\eta,y) = a(x, \widetilde{M}(x,y)\eta)\, \chi(x-y)\, |\det \widetilde{M}(x,y)|.$$

$b(x,\xi)$ is the ordinary symbol that comes from the compound symbol $a(x,\eta,y)$. In order to show that $b(x,\xi) \in S^m_{\tilde{\rho}}$, apply the principal lemma

with $\quad \rho(x,\xi) = \rho_o(x, \widetilde{M}(x)\xi)$

and $\quad \rho(x,y,\eta) = \rho_o(x, \widetilde{M}(x,y)\eta)$.

The estimate for $b(x,\xi)$ undifferentiated comes from the choices for C:

$$C(x,y,\eta,\xi) = \rho_o(x, \widetilde{M}(x,y)\eta)^m \quad \text{and} \quad C(x,\xi) = \rho_o(x, \widetilde{M}(x)\xi)^m.$$

We omit the straightforward details.

Remark. Note the similarity with Proposition 1, Section 4. In the applications in Chapter IV, we will also represent our operators in the form of Theorem 18.

Let $\psi: \Omega_1 \longrightarrow \Omega_2$, be a diffeomorphism of open subsets of $\mathbb{R}^n$. Define $T_\psi f(x) = f(\psi(x))$. Let $\widetilde{J}_{\psi(x)}$ denote the contragredient of the Jacobian matrix J_x of ψ at x, i.e., $\widetilde{J}_{\psi(x)} = {}^tJ_x^{-1}$. We want a theorem that says roughly that

$$T_\psi \circ a(x,D) = b(x,D) \circ T_\psi$$

where $b(x,\xi) \in S^m_{\widetilde{\rho}}$, with $\widetilde{\rho}(x,\xi) = \rho(\psi(x), \widetilde{J}_{\psi(x)}\xi)$. A precise statement requires cut-off functions η_1 and $\eta_2 \in C^\infty(\Omega_2)$.

Corollary. If $a(x,\xi) \in S^m_\rho$, then

$$T_\psi \eta_1 a(x,D)\eta_2 = \eta_2' b(x,D)\eta_1' T_\psi$$

with b as above (and $\eta_2', \eta_1' \in C_0^\infty(\Omega_1)$).

Proof. $T_\psi \eta_1 a(x,D)\eta_2 f(x) =$

$$= \iint \eta_2(\psi(x))\, a(\psi(x),\xi)\, e^{-2\pi i(y-\psi(x))\cdot\xi}\, \eta_1(y) f(y)\, dy\, d\xi =$$

$$= \iint \eta_2(\psi(x))\, a(\psi(x),\xi)\, e^{2\pi i(\psi(y)-\psi(x))\cdot\xi}\, \eta_1(\psi(y)) f(\psi(y))\, |\det J_y|\, dy\, d\xi$$

Now apply Theorem 18 with $\Phi(x,y) = \psi(y) - \Psi(x)$.

The corollary implies that it is natural to define ρ as a function on the cotangent bundle $T^*(M)$ of a compact manifold M. (The conditions on φ make sense in this context because $d\varphi$ is interpreted as $d\varphi \in T^*(M)$.)

Furthermore, the symbol class S^m_ρ is invariantly defined on $T^*(M)$ for such ρ, although each individual symbol is not. To do this, choose a partition of unity $\{\varphi_j\}$ so fine that when $\operatorname{supp}\varphi_j \cap \operatorname{supp}\varphi_k \neq \emptyset$, $\operatorname{supp}\varphi_j \cup \operatorname{supp}\varphi_k$ is contained in a coordinate chart, so that $\varphi_k T\varphi_j \in S^m_\rho$ makes sense locally. An operator $T = \sum_{j,k} \varphi_k T\varphi_j$ corresponds to the class S^m_ρ if

$$\varphi_k T\varphi_j \in S^m_\rho \qquad \text{whenever } \operatorname{supp}\varphi_j \cap \operatorname{supp}\varphi_k \neq \emptyset$$

and $\quad \varphi_k T\varphi_j$ has C^∞ kernel $\quad$ whenever $\operatorname{supp}\varphi_j \cap \operatorname{supp}\varphi_k = \emptyset$

(The reader may check that in view of the corollary this definition is invariant under the choice of coordinate charts.)

We mention one last property of our symbols, a restriction property. Suppose that $x_1 \in \mathbb{R}^{n_1}$, $x_2 \in \mathbb{R}^{n_2}$, $x = (x_1, x_2) \in \mathbb{R}^n = \mathbb{R}^{n_1} \times \mathbb{R}^{n_2}$, and similarly for $\xi = (\xi_1, \xi_2)$. Suppose f is a function of x_1, g is a function of (x_1, x_2), and $\psi \in C_0^\infty(\mathbb{R}^{n_2})$ with $\psi \equiv 1$ for x_2 near 0. Denote

$$(Ef)(x_1, x_2) = f(x_1)\psi(x_2)$$

$$(Rg)(x_1) = g(x_1, 0)$$

Proposition 5. $R \circ a(x, D) \circ E = b(x_1, D_1)$, where $b(x_1, \xi_1) \in S^m_{\rho_1}$ and $\rho_1(x_1, \xi_1) = \rho(x_1, 0, \xi_1, 0)$.

We shall see that up to an $S^{-\infty}$ error, $b(x_1, \xi_1) = a(x_1, 0, \xi_1, 0)$. (Clearly $a(x_1, 0, \xi_1, 0) \in S^m_{\rho_1}$.) In fact, $R \circ a(x, D) \circ Ef(x_1) =$

$$= \iint a(x_1, 0, \xi_1, \xi_2)\, e^{2\pi i(y_1 - x_1)\cdot\xi_1}\, e^{2\pi i y_2\cdot\xi_2} f(y_1)\psi(y_2)\, dy\, d\xi$$

$$= \iint a(x_1, 0, \xi_1, \xi_2) e^{2\pi i (y_1 - x_1)\cdot\xi_1} \hat{\psi}(\xi_2) f(y_1)\, dy_1 d\xi$$

Therefore

$$b(x_1, \xi_1) = \int a(x_1, 0, \xi_1, \xi_2) \hat{\psi}(\xi_2)\, d\xi_2$$

Notice that $\hat{\psi}_2$ has the properties

$$\int \hat{\psi}(\xi_2)\, d\xi_2 = 1, \quad \int \hat{\psi}(\xi_2)\, \xi_2^{\alpha}\, d\xi_2 = 0 \quad \text{for } \alpha \neq 0.$$

By Taylor's theorem,

$$a(x_1, 0, \xi_1, \xi_2) = a(x_1, 0, \xi_1, 0) + \sum_{|\alpha| \leq N-1} a^{\alpha}(x_1, 0, \xi_1, 0)\, \xi_2^{\alpha} + R_N .$$

Thus $b(x_1, \xi_1)$ differs from $a(x_1, 0, \xi_1, 0)$ by a remainder term $\int R_N \hat{\psi}(\xi_2)\, d\xi_2$ that is not hard to estimate.

Warning. In the case when φ is not constant, ρ_1 need not be an admissible ρ function. For example, if $\varphi(x, y) = y - x^2$, then $\varphi(x, 0) = -x^2$ vanishes to too high order at $x = 0$. ρ_1 will be an admissible ρ function exactly when the surface of restriction is transverse to the zero set of φ. Fortunately, we will only need Proposition 5 in the case when φ is constant.

Chapter IV. Applications

We shall now present the main applications of our theory. Our first example arises in the setting of a domain $\mathcal{D}$ in $\mathbb{C}^{n+1}$ with smooth strictly pseudo-convex boundary. We intend to show that the Cauchy-Szegö projection operator (as well as the related Henkin-Ramirez operators) are given by operators with symbols in the class S^0_ρ, for appropriate ρ. Similarly, parametricies for $\Box_b$ (acting on q-forms, where $0<q<n$) are given in terms of symbols of class S^{-2}_ρ. The proof of these facts proceeds as follows. First, natural coordinates are introduced at each point of the boundary of $\mathcal{D}$ in terms of an approximation by the Heisenberg group (see §12). This allows us to write the required operators as singular integral operators of "type λ" in the sense of Folland and Stein [16]. Then one proves, by appealing to Theorem 18, that such operators correspond to symbols of class S^{-2}_ρ. The same methods can be used to treat parametricies of a class of operators of Hörmander type (see § 14). In §15 we deal with the oblique derivative problem which requires some further ideas, and finally in §16 we discuss another important example, first treated by Kannai.

§12. Normal coordinates for pseudoconvex domains

Let $r(x)$ be a defining function for a domain $\mathcal{D} \subseteq \mathbb{C}^{n+1}$. That is, $r \in C^\infty$, $r(x) = 0$ implies $dr(x) \neq 0$.

$$\mathcal{D} = \{x \in \mathbb{C}^{n+1} : r(x) < 0\}$$

$$b\mathcal{D} = \{x \in \mathbb{C}^{n+1} : r(x) = 0\}$$

The tangent space T_x to $b\mathcal{D}$ at x has (real) dimension $2n+1$. It has an intrinsic subspace $\widetilde{T}_x$ of dimension $2n$ defined by $\widetilde{T}_x = \{v: v \in T_x$ and $iv \in T_x\}$. (Another way to obtain $\widetilde{T}_x$ is to take real and imaginary parts of all holomorphic vector fields $\sum_{j=1}^{n+1} a_j \frac{\partial}{\partial z_j}$ tangent to $b\mathcal{D}$ at x.) Dual to T_x is the cotangent space T_x^*. It has a natural one-dimensional subspace $\widetilde{T}_x^*$, the annihilator of $\widetilde{T}_x$.

Let Q_2 be a uniformly positive definite quadratic form on $T^*(b\mathcal{D})$. Let Q^1 be a positive semi-definite form on $T^*(b\mathcal{D})$ that can be written as the sum of squares of linear functionals on T_x^* smoothly varying in x, and with joint null space exactly $\widetilde{T}_x^*$. The natural ρ function on $b\mathcal{D}$ is

(1) $$\rho(x,\xi) = (Q_x^1(\xi)^2 + Q_2(\xi))^{1/4}$$

We will assume that $\mathcal{D}$ is strictly pseudoconvex. Strict pseudoconvexity will enable us to approximate $b\mathcal{D}$ locally with the Heisenberg group. Recall that $\mathbb{H}^n = \{\zeta \in \mathbb{C}^n,\ t \in \mathbb{R}\}$ was identified with $b\mathcal{D}_0$ with $\mathcal{D}_0 = \{z \in \mathbb{C}^{n+1}: \operatorname{Im} z_{n+1} > |z_1|^2 + \dots + |z_n|^2\}$. If $\mathcal{U}_0$ is a neighborhood of $y \in b\mathcal{D}_0$, define the mapping $\Theta_0: \mathcal{U}_0 \times \mathcal{U}_0 \longrightarrow \mathbb{H}^n$ by $\Theta_0(x,y) = y^{-1}x$. We wish to construct a similar mapping

$$\Theta : \mathcal{U} \times \mathcal{U} \longrightarrow \mathbb{H}^n$$

where $\mathcal{U}$ is a neighborhood of $y \in b\mathcal{D}$ and

(a) $x \longmapsto \Theta(x,y)$ is a diffeomorphism of $\mathcal{U}$ to a neighborhood of 0 in $\mathbb{H}^n$.

(b) $\Theta(y,y) = 0$

Further properties of Θ are stated below.

The left invariant vector fields on $\mathbb{H}^n$, Z_j, $\overline{Z}_j$, and T satisfy

(2) $$[\overline{Z}_j, Z_k] = 2i\,\delta_{jk}T \qquad j=1,\ldots,n,\ \ k=1,\ldots,n$$

and all other commutators vanish. Choose a basis for $T(b\mathcal{D})$, $W_1,\ldots,W_n$, $\overline{W}_1,\ldots,\overline{W}_n$, S, where S is real and $W_1,\ldots,W_n$ are holomorphic vector fields (i.e., linear combinations of $\frac{\partial}{\partial z_1},\ldots,\frac{\partial}{\partial z_{n+1}}$) and $\overline{W}_1,\ldots,\overline{W}_n$ are anti-holomorphic vector fields, which are tangential. We want $W, \overline{W}$ to resemble $Z, \overline{Z}$ as much as possible.

Think of $W, \overline{W}$ as order 1 and S of order 2. $[W_j, W_k]$ is a (tangential) holomorphic vector field, and hence a linear combination of $W_1,\ldots,W_n$. This is one order lower than one might expect, so in some sense $[W_j, W_k]$ vanishes modulo error terms. Similarly, $[\overline{W}_j, \overline{W}_k]$, $[W_j, S]$, and $[W_j, S]$ vanish modulo admissible error terms. This leads us to

(3) $$[\overline{W}_j, W_k] = 2i\,a_{jk}S + \text{linear comb. of } W, \overline{W}$$

Clearly (a_{jk}) is a Hermitian matrix. Furthermore, for appropriate choice of sign of S, $b\mathcal{D}$ is strictly pseudoconvex if and only if (a_{jk}) is (strictly) positive definite. This is independent of basis $\{W_j, \overline{W}_l, S\}$ chosen. (See Folland-Kohn [15], p. 56.) Passing to another appropriate basis changes (3) to

(4) $$[\overline{W}_j, W_k] = 2i\,\delta_{jk}S \qquad \text{mod } (W, \overline{W})$$

The additional properties of Θ are (if (4) is satisfied):

(c) Fix y. In coordinates assigned by the mapping $x \longrightarrow \Theta(x,y) = (\zeta, t) \in \mathbb{H}^n$,

$$W_j = Z_j + O^1 \frac{\partial}{\partial \zeta_j} + O^1 \frac{\partial}{\partial \bar{\zeta}_j} + O^2 \frac{\partial}{\partial t}$$

$$\overline{W}_j = \overline{Z}_j + \text{same error}$$

(O^k means vanishing like $O(|\zeta| + |t|^{1/2})^k$ as $(\zeta, t) \to 0$.)

(d) $\Theta(x, y) = -\Theta(y, x)$. [We will not need this fact here. In the second construction below equality only holds modulo an error term.]

We will outline two different constructions of Θ. They appear in Folland-Stein [16] §14 and §18. (See also Greiner-Stein [20] Chapter 4 and Rothschild-Stein [39].)

Construction 1: The exponential mapping

Suppose X is a smooth real vector field, then $\exp(tX) = \varphi_t$ is a family of diffeomorphisms for $-\varepsilon < t < \varepsilon$ such that

$$\varphi_t \circ \varphi_s = \varphi_{t+s};\ \varphi_0 = \text{Id}.$$

and
$$\frac{\partial}{\partial t} f(\varphi_t(x))\big|_{t=0} = Xf(x).$$

If $X_1, \ldots, X_n$ *are smooth, linearly independent vector fields on a manifold* M of dimension n, then $(t_1, \ldots, t_n) \to \exp(t_1 X_1 + \ldots + t_n X_n)(x_0)$ is a diffeomorphism of some small neighborhood of 0 in $\mathbb{R}^n$ to a neighborhood of x_0 in M.

The coordinates (ζ, t) on $\mathbb{H}^n$ are induced by the exponential mapping

$$(x_1, \ldots, x_n, y_1, \ldots, y_n, t) \to \exp(x_1 X_1 + \ldots + x_n X_n + y_1 Y_1 + \ldots + y_n Y_n + tT)\ (0)$$

with $X_j = 2\,\mathrm{Re}\, Z_j$, $Y_j = -2\,\mathrm{Im}\, Z_j$. It is therefore appropriate to assign coordinates $\Theta(x, y)$ to a point x in U by the (inverse) exponential mapping based on $2\operatorname{Re} W_j$, $2\operatorname{Im} W_j$, S originating at the point y. It is not hard to show that this function Θ does satisfy (a) - (d).

Construction 2. The Levi polynomial

Suppose $y \in b\mathcal{D}$ and $\mathcal{D}$ is strictly pseudoconvex. The Levi polynomial is a polynomial $P_y(x)$ (of degree ≤ 2) such that locally $\{x: P_y(x) = 0\}$ lies outside $\mathcal{D}$ and has contact at y of order 2. (It then follows that $f(x) = 1/P_y(x)$ is holomorphic in $\mathcal{D}$ near y but cannot be extended across y.)

Our second method for construction of Θ is the same as the method for construction of $P_y(x)$. In the case

$$\mathcal{D}_0 = \{z = \operatorname{Im} z_{n+1} > |z_1|^2 + \ldots + |z_n|^2\}$$

the Levi polynomial at 0 is $P_0(z) = z_{n+1}$. In general, if $\mathcal{D} = \{x: r(x) < 0\}$, for appropriate choice of holomorphic coordinates $z_1, \ldots, z_{n+1}$.

$$r(x) = -\operatorname{Im} z_{n+1} + \sum_{j=1}^{h} |z_j|^2 + \text{Error} \tag{5}$$

where Error $= O^3 = O(|z_1| + \ldots + |z_n| + |z_{n+1}|^{1/2})^3$ as $x \longrightarrow y$. This can be done with a linear change of coordinates followed by a quadratic change (see [16], §18). (The choice of coordinates depends smoothly on y.)

In the new coordinates, $P_y(x) = z_{n+1}$ is the Levi polynomial. In coordinates $z_1, \ldots, z_{n+1}$, $\mathcal{D}$ looks very much like $\mathcal{D}_0$. Thus we define

$$\Theta(x,y) = (\zeta, t) \quad \text{where } \zeta = (z_1, \ldots, z_n),\ t = \operatorname{Re} z_{n+1}.$$

Again it is possible to verify (c), and (d) holds up to an error.

§13. $\Box_b$ and the Cauchy-Szegö integral

Our pseudo-differential operators will be represented as "operators of type λ" in the sense of [16]. The basic building block for an operator of type $\lambda \geq 0$ is an operator

$$f \mapsto \int a(x)\, k(\Theta(x,y))\, b(y)\, f(y)\, dy$$

where dy is any smooth measure that is positive everywhere, $a, b \in C_0^\infty$, and k is a distribution which is C^∞ away from the origin and homogeneous of degree $-2n-2+\lambda$ on $\mathbb{H}^n$. (See Theorem 1, Sections 1 and 3. The homogeneous dimension of $\mathbb{H}^n$ is $2n+2$.) Note that when $\lambda = 0$, k must be the sums of a function of mean-value zero and a multiple of the delta function at the origin. The integral above exists in a principal value sense as

$$c(x)\, f(x) + \lim_{\epsilon \to 0} \int_{|\Theta(x,y)| > \epsilon} a(x)\, k(\Theta)\, (x,y))\, b(y)\, f(y)\, dy$$

$|\cdot|$ denotes $\mathbb{H}^n$ norm.

<u>Definition.</u> An operator $T = f \longrightarrow \int k(x,y)\, f(y)\, dy$ is of type λ if for each $m > 0$ we can write $K(x,y) = \sum_{j=1}^{M} a_j(x)\, k_j(\Theta(x,y)) b_j(y) + E_m(x,y)$ where $E_m \in C_0^{(m)}$, $a_j, b_j \in C_0^\infty$ and k_j is a distribution of degree $-2n-2+\lambda_j$ on $\mathbb{H}^n$ with $\lambda_j \geq \lambda$, C^∞ away from the origin. We will call the kernel $K(x,y)$ a kernel of type $-2n-2+\lambda$.

<u>Theorem 19.</u> <u>Every operator of type</u> λ <u>is a pseudo-differential operator</u>

with symbol in $S_\rho^{-\lambda}$. (ρ was defined in (1).)

Proof. It is not hard to reduce matters to the case of an operator T with kernel $a(x)\,K(\Theta(x,y))\,b(y)$, where $a(x)$ and $b(y)$ vanish outside of a coordinate patch where $\Theta(x,y)$ is defined. Choose $\psi \in C_0^\infty(\mathbb{H}^n)$ so that $\psi \equiv 1$ in a neighborhood of the origin. Let $K_0(\zeta,t) = k(\zeta,t)\,\psi(\zeta,t)$. We may as well assume that

$$a(x)\,k(\Theta(x,y))\,b(y) = a(x)\,K_0(\Theta(x,y))\,b(y).$$

Let $\xi_1,\ldots,\xi_n$ be dual to $\operatorname{Re}\zeta$, $\xi_{n+1},\ldots,\xi_{2n}$ dual to $\operatorname{Im}\zeta$, ξ_{2n-1} dual to t. (The natural dilations on ξ are $r\xi = (r\xi_1,\ldots,r\xi_{2n}, r^2\xi_{2n+1})$.)

Obviously $\hat{K}_0(\xi) \in C^\infty$. Moreover, $\hat{K}_0(\xi)$ is the sum of a function in $\mathcal{S}$ with a function that is homogeneous of degree $-\lambda$ for large ξ. (This follows from the proof of Theorem 1.) Define

$$\rho_0(\xi) = \left(\left(\sum_{j=1}^{2n} \xi_j^2\right)^2 + \|\xi\|^2\right)^{1/4}.$$

A simple homogeneity argument shows that $\hat{K}_0(\xi) \in S_{\rho_0}^{-\lambda}$.

Using local coordinates on $\mathbb{R}^n$

$$Tf(x) = \int_{\mathbb{R}^n} a(x)\,K_0(\Theta(x,y))\,b(y)\,f(y)\,dy$$

(For a suitable modification of $b(y)$ we can choose dy to be Lebesgue measure.) Now notice that

$$K_0(\Theta(x,y)) = \int e^{-2\pi i\,\Theta(x,y)\cdot\xi}\,\hat{K}_0(\xi)\,d\xi.$$

Therefore,

$$Tf(x) = \iint a(x)\, e^{2\pi i \Phi(x,y)\xi}\, \hat{K}_0(\xi)\, b(y)\, f(y)\, dy\, d\xi\,,$$

where $\Phi(x,y) = -\Theta(x,y)$.

Theorem 18 implies that T has a symbol in the class $S_\rho^{-\lambda}$ where $\rho(y,\xi) = \rho_0(\widetilde{M}(y)\xi)$ with

$$M(y) = -\nabla_x \Phi(x,y)\big|_{x=y}{}^{*}$$

It remains to be seen that the function ρ defined here is of the same form as the one defined in (1). In fact,

$$M(y)\colon T_y(b\mathcal{D}) \xrightarrow{\sim} T_0(\mathbb{H}^n).$$

The error in the correspondence between $W_1, \ldots, W_n, \overline{W}_1, \ldots, \overline{W}_n$ and $Z_1, \ldots, Z_n, \overline{Z}_1, \ldots, \overline{Z}_n$ vanishes as x tends to y. Therefore,

$$M(y) = \widetilde{T}_y(b\mathcal{D}) \xrightarrow{\sim} \widetilde{T}_0(\mathbb{H}^n)$$

whence, taking the transpose,

$${}^tM(y) = \widetilde{T}_0^*(\mathbb{H}^n) \xrightarrow{\sim} \widetilde{T}_y^*(b\mathcal{D})$$

or

$$\widetilde{M}(y)\colon \widetilde{T}_y^*(b\mathcal{D}) \xrightarrow{\sim} \widetilde{T}_0^*(\mathbb{H}^n).$$

Since $\widetilde{T}_0^*(\mathbb{H}^n) = \{(0, \xi_{2n+1})\colon \xi_{2n+1} \in \mathbb{R}\}$ is the null space of $\sum_{j=1}^{2n} \xi_j^2$, ρ has the desired form. qed.

<u>Corollary</u>. <u>When $\Box_b$ is restricted to</u> q-forms <u>with</u> $0 < q < n$, <u>it has a parametrix which is an operator with symbol in the class</u> S_ρ^{-2}.

---

* In Theorem 18 we used the matrix $M(x) = \nabla_y \Phi(x,y)\big|_{y=x}$, but one easily sees that the roles of x and y are reversible in the proof of the Observation, in Section 11.

This follows from the fact (see Folland-Stein [16]) that a parametrix is an operator of type 2.

It is to be noted that in order to use the results of [16] we require that the underlying Hermitian metric in terms of which $\Box_b$ is defined to be a "Levi-metric," in the sense of [20], §13. This condition can be dropped, as well as the restriction that $\mathcal{D}$ be strictly pseudoconvex (and replaced by the condition $Y(q)$), as we shall see in §14.

Theorem 20. The Szegö projection on a strictly pseudoconvex domain has a symbol of class S^0_ρ (ρ as above).

Proof. Let $r(x)$ be a defining function for $\mathcal{D}$. C. Fefferman's [13] asymptotic formula for the Szegö kernel $S(x,y)$ (in the form due to Boutet de Monvel-Sjöstrand [6]) is

$$S(x,y) = A(x,y)\,\psi(x,y)^{-n-1} + B(x,y)\log\psi(x,y)$$

where A, B are smooth in x and y jointly and

$$\psi(x,y) = \overline{\psi(y,x)}$$

$\psi(u,u) = r(u)$ for $u \in \overline{\mathcal{D}} \times \overline{\mathcal{D}}$, and more generally

$$(6) \qquad \psi(u+x,u+y) \sim \sum \frac{\partial^{\alpha+\beta} r(u)}{\partial^{\alpha}u\,\partial^{\beta}\overline{u}}\,\frac{x^{\alpha}}{\alpha!}\,\frac{\overline{y}^{\beta}}{\beta!}$$

For example, in the unit ball,

$$r(x) = 1 - |x|^2$$

$$S(x,y) = c_n/(1-x\cdot\overline{y})^{n+1}$$

Fact 1. $A(x,y)\psi(x,y)^{-n-1}$ is a kernel of type $-2n-2$ and hence corresponds to an operator with symbol in S^0_ρ

Fact 2. $B(x,y)\log\psi(x,y)$ is very nearly a kernel of type $2n+2$. Precisely, it corresponds to a symbol of the class $S_\rho^{-2n-2+\epsilon}$ for any $\epsilon > 0$.

In order to prove these facts, we use the second construction of Θ. We are given coordinates $z_1, \ldots, z_{n+1}$ with

$$\Theta(x,y) = (\zeta, t), \quad \zeta = (z_1, \ldots, z_n), \quad t = \operatorname{Re} z_{n+1}$$

and
$$r(x) = -\operatorname{Im} z_{n+1} + \sum_{j=1}^{n} |z_j|^2 + \text{Error}.$$

Using (6), one can calculate (see Phong-Stein [37])
$\psi(x,y) = -iz_{n+1} + O^3 = |\zeta|^2 - it + O^3$. Hence,

$$(\psi(x,y))^{-n-1} = (|\zeta|^2 - it + O^3)^{-n-1} = (|\zeta|^2 - it)^{n-1}\left(1 + \frac{O^3}{|\zeta|^2 - it}\right)^{-n-1}$$
$$= (|\zeta|^2 - it)^{-n-1} \sum_{k=0}^{N} C_k \frac{O^{3k}}{(|\zeta|^2 - it)^k} + C^{(m)} \text{ Error}.$$

Now expand O^{3k} in a Taylor series in ζ and t to see that we have a kernel of type $-2n-2$. (Notice that the leading term $(|\zeta|^2 - it)^{-n-1}$ does have mean-value zero.) To see that A is harmless, fix y and expand in a Taylor series in x (with (ζ, t) coordinates).

$$A(x,y) = \sum_{|\alpha| \leqq N} a_\alpha(y)(\zeta, t)^\alpha + \text{Error}.$$

For Fact 2, the same sort of reduction leads us to the kernel $\log(|\zeta|^2 - it)$ on $\mathbb{H}^n$ (cut-off for large (ζ, t)). A fairly easy calculation shows that

$$[\text{cut-off} \log(|\zeta|^2 - it)]^\wedge \simeq c_1 \xi_{2n+1}^{-n-1} e^{c_2 |\xi'|^2/\xi_{2n+1}} \quad \text{for large } \xi .$$

This shows that the kernel has a symbol of class S_ρ^{-2n-2}. In general, estimates are not quite as precise. The growth of $B(x,y) \log \psi(x,y)$ and its derivatives are dominated by that of $\psi(x,y)^{-\varepsilon/2}$, so we get a symbol of class $S_\rho^{-2n-2+\varepsilon}$. Thus the log term is negligible compared to the leading term $A(x,y)(\psi(x,y))^{-n-1}$.

Corollary. The Henkin-Ramirez projection operators correspond to symbols of the class S_ρ^0

This follows because these operators are integral operators of type 0. See Kerzman and Stein [27] for further details about this point.

For the operators discussed above both the isotropic and non-isotropic Sobolev space and Hölder estimates hold. This is a consequence of the following fact.

Proposition 1. For the ρ function used above (given by (1)), the assumption A_1 of Section 9, Chapter 3 holds.

The operator $\mathcal{L}_0 = -\frac{1}{2} \sum_{s=1}^{n} (W_j \overline{W}_j + \overline{W}_j W_j)$ has a (right) parametrix A of type 2, with error of type 1, see Folland-Stein [16]. Thus

$$\Sigma \left\{ W_j \left(\frac{-\overline{W}_j A}{2} \right) + \overline{W}_j \left(\frac{-W_j A}{2} \right) \right\} - \text{Id.} = E$$

where we may take $\{P_k\} = \{W_j, \overline{W}_j\}$ and $-Q_k = \left\{ \frac{\overline{W}_j A}{2}, \frac{W_j A}{2} \right\}$, with E having symbol of class S_ρ^{-1}, if we use Theorem 19.

§14. Operators of Hörmander and Grushin

Let Ω be an open set in $\mathbb{R}^n$. Let $X_0, X_1, \ldots, X_m$ be m+1 smooth real vector fields on Ω. Suppose that

(7) $X_0, \ldots, X_m$ and $[X_j, X_k]$, $\leq j < k \leq m$ span the tangent space at each point of Ω.

Denote $\mathcal{L} = X_0 + \sum_{j=1}^{m} X_j^2$. We want to examine the parametrix of $\mathcal{L}$. (One might hope to replace (7), which we call a step 2 hypothesis, with the general hypothesis of Hörmander that the commutators of $X_0, X_1, \ldots, X_m$ up to some finite order span the tangent space at each point.* This will require an extension of our theory; some of the relevant needed ideas are sketched in [35].)

If $X_j = \sum_{k=1}^{n} a_k^j(x) \frac{\partial}{\partial x_k}$ $\quad j=1, \ldots, m,$

then $L_j(\xi) = \sum_{k=1}^{n} a_k^j(x)\, \xi_k$ is the symbol of X_j.

Let $\quad Q_x(\xi) = \sum_{j=1}^{m} |L_j(\xi)|^2$ and $\rho(x,\xi) = (Q_x(\xi)^2 + \|\xi\|^2)^{1/4}$.

Theorem 21. *$\mathcal{L}$ has a two-sided (local) parametrix whose symbol belongs to S_ρ^{-2}. More precisely, for each $\eta \in C_0^\infty(\Omega)$ there exists $a \in S_\rho^{-2}$ such that*

$$a(x,D)\,\mathcal{L} = \eta\, \text{Id.} + E_1$$

$$\mathcal{L}\, a(x,D) = \eta\, \text{Id.} + E_2$$

where $E_j = e_j(x,D)$, $e_j \in S_\rho^{-k}$ for any fixed k.

---

*See Hörmander [23a].

Proof. We wish to follow a procedure like the one above where we approximated $b\mathcal{D}$ by $\mathbb{H}^n$. But the vector fields may not be linearly independent. Therefore, we extend the vector fields $X_0, X_1, \ldots, X_m$ to vector fields $\widetilde{X}_0, \ldots, \widetilde{X}_m$ on $\Omega \times \mathbb{R}^M$ (possibly shrinking Ω) such that for $x \in \Omega$, $y \in \mathbb{R}^M$,

(8) $\widetilde{X}_j = X_j + \sum_\ell a_\ell(x,y) \frac{\partial}{\partial y_\ell} \qquad j=0,\ldots,m$

(9) $\widetilde{X}_j$, $j=0,1,\ldots,m$ and $[\widetilde{X}_j, \widetilde{X}_k]$, $1 \leq j < k \leq m$, are linearly independent at each point.

(10) The vector fields in (9) span the tangent space to $\Omega \times \mathbb{R}^M$ at each point. Notice that M is determined. In fact,

$$n + M = \#\text{ linear independent vectors} = m+1 + \frac{m(m-1)}{2}.$$

To achieve (8), (9), and (10) is not difficult. We leave it as good an exercise for the interested reader. When higher commutators are involved, the proof of the corresponding result is more difficult (see Rothschild-Stein [39]).

We are now in a position to approximate $\Omega \times \mathbb{R}^M$ by the Lie group G where Lie algebra is spanned by

$$Y_0, \ldots, Y_m, \; Y_{j,k} \quad 1 \leq j \leq k \leq m$$

with relations

$$[Y_j, Y_k] = Y_{j,k}$$

and all other commutators vanish.

Choose the dilation structure that multiplies $Y_1, \ldots, Y_m$ by δ and Y_0,

$Y_{j,k}$ by δ^2. This makes the operator $\bar{\mathcal{L}} = Y_0 + \sum_{j=1}^{m} Y_j^2$ homogeneous of degree 2.

Let a be the homogeneous dimension of G. (One easily sees that $a = m + 2 + (m)(m-1) = m^2 + 2$.) By Hörmander's theorem $\bar{\mathcal{L}}$ is hypoelliptic. One can use a homogeneity argument to show that $\bar{\mathcal{L}}$ has a unique homogeneous fundamental solution k (which must be a distribution on G of class $-a+2$).

Define Θ as in our first construction using the exponential mapping so that $\tilde{X}_j$, $[\tilde{X}_j, \tilde{X}_k]$ correspond to Y_j, $Y_{j,k}$. (This is where we need linear independence.) $k(\Theta(x,y))$ will be the kernel of a parametrix for $\tilde{X}_0 + \sum_{j=1}^{m} \tilde{X}_j^2$.

Up to this point the argument is taken from Rothschild-Stein [39], which should be consulted for further details.

At this stage we can invoke again Theorem 18, to prove the analogue of Theorem 19 in our setup. Thus the parametrix will have a symbol in the class $S_{\tilde{\rho}}^{-2}$, where $\tilde{\rho}$ is associated to $\tilde{X}_j$ in the same way that ρ was associated to X_j. Now the restriction theorem (Proposition 5, §11) yields a parametrix for $\mathcal{L}$ with symbol in the class S_ρ^{-2}.

Next suppose we are given a smooth domain $\mathcal{D}$ in $\mathbb{C}^{n+1}$ whose boundary satisfies the condition Y(q) in the sense of Folland-Kohn [15], namely if at each point of the boundary of $\mathcal{D}$, if p_1 is the larger of the number of eigenvalues of the same sign of the Levi form, and p_2 is the number of pairs of eigenvalues of opposite sign, then

(11) $\qquad p_1 \geq \max(q+1, n+1-q)$ or $p_2 \geq \min(q+1, n+1-q)$.

By the same method of proof as that of Theorem 21, (using here the construction in §19 of [39]) we obtain

Corollary. When $\Box_b$ is restricted to q-forms and condition Y(q) is satisfied, then $\Box_b$ has a parametrix which is an operator with symbol in the class S_ρ^{-2}. Here ρ is given by (1) in §12. In addition it is no longer required that the Hermitian metric used be a Levi metric.

We shall now give a more explicit description of the parametrix of operators of the Grushin type: $\frac{\partial^2}{\partial t^2} + t^2 \frac{\partial^2}{\partial x^2}$, and their variants.* More precisely, let α be a complex number and set

(12) $$\mathcal{L}_\alpha = \frac{\partial^2}{\partial t^2} + t^2 \frac{\partial^2}{\partial x^2} + i\alpha \frac{\partial}{\partial x}$$

When $\alpha \neq \pm 1, \pm 3, \pm 5, \ldots$ we shall exhibit the exact symbol of an operator $\mathcal{P}_\alpha$ so that

(13) $$\mathcal{P}_\alpha \mathcal{L}_\alpha = \mathcal{L}_\alpha \mathcal{P}_\alpha = I$$

In addition when $k = \pm$ an odd integer, we shall exhibit the symbols of the operators $\widetilde{\mathcal{P}}_k$ and Q_k with

(13′) $$\widetilde{\mathcal{P}}_k \mathcal{L}_k = \mathcal{L}_k \widetilde{\mathcal{P}}_k = I - Q_k$$

where Q_k is the orthogonal projection on the L^2 null-space of $\mathcal{L}_k$.

One way of approaching this problem (at least when $\alpha = 0$) is to reexamine the proof of Theorem 21, if we take $X_1 = \frac{\partial}{\partial t}$, and $X_2 = t\frac{\partial}{\partial x}$. If we also observe that X_1 and X_2 generate the three-dimensional

*See Grushin [20a].

Heisenberg Lie algebra, and use the results quoted in §4 then we can also obtain the formulas for the symbols of $\mathcal{P}_\alpha$, when $\pm\alpha$ is not an odd integer. However, we shall not pursue this line of reasoning, but instead we shall describe a more elementary way of finding the symbols of $\mathcal{P}_\alpha$, $\tilde{\mathcal{P}}_k$ and Q_k.

We consider instead of (12), the ordinary differential operator L_α given by

$$L_\alpha = \frac{d^2}{dt^2} - \lambda^2 t^2 + \alpha\lambda, \tag{14}$$

where λ is fixed and positive, and where we assume to begin with that $\mathrm{Re}(\alpha) < 1$. We study the inverse of (14) and its symbol by the use of the Hermite functions, $H_n(t)$, defined by

$$\{2^n n! \sqrt{\pi}\}^{-1/2} (-1)^n e^{t^2/2} \frac{d^n(e^{-t^2})}{dt^n} \qquad n=0, 1, 2, \ldots$$

These form an orthonormal basis of $L^2(-\infty, \infty, dt)$, and H_n is an eigen-function of $\frac{d^2}{dt^2} - t^2$ with eigenvalue $-(2n+1)$. (For this, and further facts we whall need about the Hermite expansion one may consult Wiener[45], Chapter I.)

If we set $H_n^\lambda = H_n^\lambda(t) = \lambda^{1/4} H_n(\lambda^{1/2} t)$ with $\lambda > 0$, we see that the H_n^λ are also an orthonormal basis for $L^2(-\infty, \infty, dt)$. Moreover, H_n^λ are eigen-functions of L_α of eigenvalue $\lambda(\alpha - 2n - 1)$. Thus the inverse of L_α has as kernel

$$\sum_{n=0}^{\infty} \frac{\lambda^{1/2} H_n(\lambda^{1/2} t) H_n(\lambda^{1/2} s)}{\lambda(\alpha - 2n - 1)}, \tag{15}$$

where the series converges in the L^2 norm (in the t-variable) after integrating against an arbitrary L^2 function in the s-variable.

The symbol of the operator whose kernel is given by (15) is

$$(16)\qquad (2\pi)^{1/2}\sum_{n=0}^{\infty}(-i)^n \frac{H_n(\lambda^{1/2}t)\,H_n(\tau/\lambda^{1/2})}{\lambda(\alpha-2n-1)}\, e^{ist\tau},$$

where τ is the dual variable to t, in view of the fact that

$$\int_{-\infty}^{\infty} e^{-it\tau} H_n(t)\,dt = (2\pi)^{1/2}(-i)^n H_n(\tau).$$

Now let $M(r,t,s)$ denote the Mehler kernel $\sum r^n H_n(t)\,H_n(s)$, $|r|<1$, which has the explicit form

$$(17)\qquad M(r,t,s) = (\pi(1-r^2))^{-1/2}\exp\left\{\frac{4tsr-(t^2+s^2)(1+r^2)}{2(1-r^2)}\right\}.$$

It is easy to see that (16) equals

$$\frac{-(2\pi)^{1/2}}{2\lambda}\, e^{it\tau}\int_0^1 r^{-1/2-\alpha/2} M(r/i,\, \lambda^{1/2}t,\, \lambda^{-1/2}\tau)\,dr,$$

and so

$$(18)\quad p_\alpha(t,\tau,\lambda) = -\sqrt{2}\,\frac{e^{it\tau}}{2\lambda}\int_0^1 r^{-1/2-\alpha/2}(1+r^2)^{-1/2}E(r)\,dr$$

with

$$E(r) = \exp\left\{\frac{-4it\tau\lambda-(\lambda^2t^2+\tau^2)(1-r^2)}{2\lambda(1+r^2)}\right\}.$$

Next, it is not difficult to see that p_α is a meromorphic function of α, with only simple poles at $\alpha = 1, 3, 5, \ldots$.

This follows by rewriting (18) when $\alpha \neq -1, -3, -5, \ldots$ as

$$(18')\qquad p_\alpha(t,\tau,\lambda) = \frac{\sqrt{2}\,e^{it\tau}}{2\lambda(e^{-i(\alpha+1)\pi}-1)}\int_L r^{-1/2-\alpha/2}(1+r^2)^{-1/2}E(r)\,dr$$

and L is a loop in the complex plane starting at 1 and passing in the upper half-plane around the origin and returning to 1 in the lower half-plane to 1.

Thus if k is an odd integer, and α is near k, we can write

$$(19) \qquad p_\alpha = \frac{(-q_k/\lambda)}{\alpha - k} + \tilde{p}_k + O(\alpha - k)$$

So $\tilde{p}_k$ is the "finite part" of p_α at $\alpha = k$.

Now if α is not an odd integer, we have

$$(20) \qquad p_\alpha(t, \frac{d}{idt}, \lambda) \circ L_\alpha = L_\alpha \circ p_\alpha(t, \frac{d}{idt}, \lambda) = I$$

and from (19) it then follows easily that

$$(20') \qquad \tilde{p}_k(t, \frac{d}{idt}, \lambda) \circ L_k = L_k \circ \tilde{p}_k(t, \frac{d}{idt}, \lambda) = I - q_k(t, \frac{d}{idt}, \lambda)$$

With these identities out of the way, we return to the Grushin operators. We let ξ be the dual variable to x, and upon taking the Fourier transform one obtains from (20)

<u>Proposition 2</u>. <u>Suppose</u> $\alpha \neq \pm 1, \pm 3, \pm 5, \ldots$. <u>Then the operator</u>

$\mathcal{L}_\alpha = \frac{\partial^2}{\partial t^2} + t^2 \frac{\partial^2}{\partial x^2} + i\alpha \frac{\partial}{\partial x}$ <u>has inverse</u> $\mathcal{P}_\alpha$ <u>whose symbol is given by</u>

$p_\alpha(t, \tau, \xi)$, <u>when</u> $\xi > 0$ <u>and</u> $p_{-\alpha}(t, \tau, -\xi)$ <u>when</u> $\xi < 0$

<u>where</u> $p_{\pm\alpha}$ <u>are given by</u> (18) <u>and</u> (18′)

In the same way when α is a positive odd integer k, then the symbol of $\tilde{\mathcal{P}}_k$ is given by $\tilde{p}_k(t, \tau, \xi)$ when $\xi > 0$, and $p_{-k}(t, \tau, -\xi)$ when $\xi < 0$. The symbol of the projection Q_k is given by $q_k(t, \tau, \xi)$ when $\xi > 0$ and 0 when $\xi < 0$. Similar results hold when $\alpha = -k$. Now if we take the

ρ function defined in terms of the quadratic form (in (τ, ξ)), $Q = \tau^2 + t^2 \xi^2$, then it is possible by elementary computations to show that the symbols of $\mathscr{P}_\alpha$ and $\mathscr{P}_k$ are of class S_ρ^{-2}, and the symbol of Q_k is of class S_ρ^0.

§15. The oblique derivative problem

The oblique derivative problem is as follows: Let $\mathscr{D}$ be a smooth region in $\mathbb{R}^{n+1}$, and let X be a given smooth real vector field given on the boundary $b\mathscr{D}$ (the "oblique derivative"). We wish to solve for u where

$$\begin{cases} \Delta u = 0 & \text{on } \mathscr{D} \\ Xu = f & \text{on } b\mathscr{D} \end{cases}$$

This question can be reduced to that of inverting a pseudo-differential operator of order one on $b\mathscr{D}$. If X is transverse to $b\mathscr{D}$, then this is essentially the classical Neumann problem, and the boundary operator is elliptic, so it can be inverted (locally); but in general this is not so.

However, if the normal component of X vanishes to order one exactly on a manifold M of codimension one in $b\mathscr{D}$, and X is transverse to M, then the symbol of the boundary operator involved is essentially

(1) $$\Omega_{\pm} = i\xi_n \pm q(x, \xi)\, x_n$$

with $q(x, \xi) = (\sum a_{ij}(x)\xi_i\xi_j)^{1/2}$, for $\|\xi\| \geq 1$

where $\{a_{ij}(x)\}$ is a smoothly varying real positive definite symmetric matrix.

Here $(x_1, \ldots, x_n)$ are appropriate local coordinates on $b\mathscr{D}$, with $M = \{x_n = 0\}$.

One can show that $i\xi_n + q(x,\xi)x_n$ can only have a right parametrix and $i\xi_n - q(x,\xi)x_n$ can only have a left parametrix. The two problems are adjoints of each other, so we shall describe in detail the results for $\Omega_+ = i\xi_n + q(x,\xi)x_n$ and its right parametrix only. For further background on this problem see Hörmander [23], and Egorov and Kondratev [11], and the literature cited there. Some more recent relevant papers are Sjöstrand [40] and Boutet de Monvel and Treves [7].

The approach to the problem of inverting Ω_+ is to replace $q(x,\xi)$ by a positive constant λ, and to try to invert the ordinary differential operator whose symbol is $i\xi_n + \lambda x_n$. Changing variables x_n to t we are led to seeking the inverse of

(2) $$D_\lambda = \frac{d}{dt} + \lambda t, \qquad \lambda > 0$$

and the symbol of this inverse.

Several words of caution need be stated. First (2) has a non-trivial (i.e., L^2) null space, namely the constant multiples of $e^{-\lambda t^2/2}$, and so only a right inverse can be found. Moreover, we need the <u>exact</u> symbol corresponding to this inverse of (2), because even if we make very good approximations (for fixed λ), these approximate symbols will not behave right under differentiation with respect to λ, and estimates of this kind are crucial in what follows.

We seek the symbol of the operator K_λ determined by

(3) $$D_\lambda K_\lambda = I,$$

and the range of K_λ is orthogonal to $e^{-\lambda t^2/2}$.

One way of obtaining this operator and its symbol is to observe that $D_\lambda D_\lambda^* = -\frac{d^2}{dt^2} + \lambda^2 t^2 + \lambda$. So $D_\lambda D_\lambda^* = -L_{-1}$ (see (14) of the previous section). Then if we use (20) we see that the symbol of K_λ is exactly the symbol of $(-D_\lambda^*) \circ p_{-1}(t, \frac{d}{idt}, \lambda)$, and so equals

$$(4) \qquad (i\tau - \lambda t)\, p_{-1}(t, \tau, \lambda) + \frac{\partial p_{-1}}{\partial t}(t, \tau, \lambda)$$

where p_{-1} is given by (18) of §14.

However, it is possible to write the symbol of K_λ in a more compact form and to give a more elementary derivation for it. This simpler derivation also applies to the higher order oblique derivative problem. (There we need to invert the operator

$$i\xi_n \pm q(x, \xi)\, x_n^k, \qquad \text{for } k > 1.$$

The results we shall derive below for the case $k = 1$ also hold for these higher values of k. See [35], where the case $k = 2$ is discussed in detail.)

Proposition 3. The operator K_λ has symbol

$$(5) \qquad \sigma(t, \tau, \lambda) = i\sqrt{2}\, e^{-\frac{1}{2}\frac{[\tau^2 + \lambda^2 t^2]}{\lambda}}\, e^{it\tau} \int_0^\infty e^{-\lambda s^2} e^{2\tau s}\, ds$$

$$- i \int_0^\infty e^{-\lambda s^2/2} e^{(\tau + it\lambda)s}\, ds, \qquad \text{if } \tau \leq 0.^*$$

Also,

$$(6) \qquad \sigma(t, -\tau, \lambda) = \overline{\sigma(t, \tau, \lambda)}.$$

Proof. We note first that if $u = K_\lambda(f)$, then $D_\lambda(u) = f$, and so $\frac{d}{dt}(e^{\lambda t^2/2} u(t)) = e^{\lambda t^2/2} f(t)$. Therefore, for some constant c_0

*The formula we stated in [34] has an error in it. The above is the correct form.

$$u(t) = \int_0^t e^{-\frac{\lambda}{2}(t^2 - z^2)} f(z)\,dz + c_0 e^{-\lambda t^2/2}.$$

Next, c_0 is determined by the assumption that $u(t)$ is orthogonal to $e^{-\lambda t^2/2}$. Carrying out the computation of c_0 we have

$$u(t) = \frac{\sqrt{\lambda}}{\sqrt{\pi}} \int_{-\infty}^{\infty} e^{-\lambda x^2} \left\{ \int_x^t e^{-\lambda(t^2 - z^2)} f(z)\,dz \right\} dx .$$

Thus

$$(7) \qquad \sigma(t, \tau, \lambda) = \frac{\sqrt{\lambda}}{\sqrt{\pi}} \int_{-\infty}^{\infty} e^{-\lambda x^2} \left\{ \int_x^t e^{-\lambda(t^2 - z^2)} e^{i(t-z)\tau} dz \right\} dx$$

From this the symmetry property (6) is obvious. Next, replace the inner integral in (7) by using contour integration in the z-variable around the rectangle (if $x < t$)

$$x \leq \mathrm{Re}(z) \leq t; \ \mathrm{Im}\, z \geq 0$$

Hence (7) leads to two integrals with the inner integrals, taken along the vertical ray $x + is$, $s \geq 0$, and along the vertical ray $t+is$, $s > 0$. So

$$(8) \qquad \sigma(t, \tau, \lambda) = \frac{i\sqrt{\lambda}}{\sqrt{\pi}} \int_{-\infty}^{\infty} e^{-\lambda x^2} \left\{ \int_0^{\infty} e^{-\frac{\lambda}{2}(t^2 - (x+is)^2)} e^{i(t-x-is)\tau} ds \right\} dx$$

$$- i\frac{\sqrt{\lambda}}{\sqrt{\pi}} \int_{-\infty}^{\infty} e^{-\lambda x^2} \left\{ \int_0^{\infty} e^{-\frac{\lambda}{2}(t^2 - (t+is)^2)} e^{i(-is)\tau} ds \right\} dx$$

The second inner integral is independent of x, and so if we carry out the x integration we get the second term of (5).

To deal with the first term of (8) requires only an evaluation of the Fourier transform of $e^{-\frac{\lambda}{2}x^2}$ at $-\lambda s - \tau$. This gives the first term of (5), and the proposition is proved.

We now define a ρ class appropriate to the symbol σ given by (8). We let ρ_0 be given by the quadratic form $Q = \tau^2 + t^2\lambda^2$ (in the (τ, λ) variables) and set $\rho_0 = \{(\tau^2+t^2\lambda^2)^2 + \tau^2 + \lambda^2\}^{1/4}$, so $\rho_0 \approx |\tau| + |t\lambda| + |\lambda|^{1/2}$. We think of (τ, λ) as the dual variable to (t, x), with everything independent of x, and with λ limited to $\lambda > 0$.

Proposition 4. $\sigma \in S^{-1}_{\rho_0}$.

To make the necessary estimates we need the following two simple lemmas.

Lemma 1. *If* m *is an integer* ≥ 1, *and* $\mathrm{Re}(z) \leq 0$, *then*

$$\left|\int_0^\infty s^{m-1} e^{-s^2/2} e^{zs}\, ds\right| \leq c_m \min\{1, |z|^{-m}\}$$

Lemma 2. *If* φ *is bounded and rapidly decreasing on* $(0, \infty)$ *then*

$$\left|\frac{t^M}{\lambda^N}\varphi\left(\frac{Q}{\lambda}\right)\right| \leq C_{M,N}\, \rho_0^{-2N-M}$$

where $Q = \tau^2 + t^2\lambda^2$, *and* $M, N \geq 0$.

The first estimate in Lemma 1 is trivial since $|e^{zs}| \leq 1$; the second follows by an m-fold integration by parts, if we use $\frac{1}{z}\frac{d}{ds}e^{zs} = e^{zs}$.

For the second lemma we divide consideration in two cases: whether $Q \leq \lambda$ or $Q > \lambda$. In the first we use the fact that $|\varphi(Q/\lambda)| \leq$ constant, and so $\frac{t^M}{\lambda^N} = \frac{(t\lambda)^M}{\lambda^{N+M}} \lesssim \frac{\rho_0^M}{\rho_0^{2N+2M}} = \rho_0^{-2N-M}$, since $\lambda \approx \rho_0^2$ in that case. In the second case we have $|\varphi(Q/\lambda)| \leq$ constant $[\frac{\lambda}{Q}]^{M+N}$, and so we get as an estimate

$$\frac{t^M}{\lambda^N}[\frac{\lambda}{Q}]^{M+N} = \frac{(t\lambda)^M}{Q^{M+N}} \leq \frac{\rho_o^M}{Q^{N+N}} \lesssim \rho_o^{-2N-M}, \text{ since } Q \approx \rho_o^2 \text{ in that case.}$$

In proving the estimates for σ it suffices to restrict attention to the region where $\tau \leq 0$, in view of the symmetry in τ given by (6); the integral representation (7) shows that σ is smooth across $\tau = 0$.

We prove first that

$$(8) \qquad |\sigma(t, \tau, \lambda)| \leq c\rho_o^{-1}.$$

For the second term in (5) we use Lemma 1, with $m = 1$, and $z = \frac{\tau + i\lambda t}{\sqrt{\lambda}}$. This gives as estimate $O(\min\{\frac{1}{\sqrt{\lambda}}, \frac{1}{|\tau + i\lambda t|}\})$ which is $O(\rho_o^{-1})$, if we consider the cases where $Q \equiv |\tau + i\lambda t|^2 > \lambda$ and $Q \leq \lambda$. For the first term in (5) we observe that $\left|\int_0^\infty e^{-\lambda s^2} e^{\tau s} ds\right| \leq c\lambda^{-1/2}$ and then use Lemma 2, with $\varphi(u) = e^{-u^2/2}$, $N = 1/2$, and $M = 0$.

The inequalities of the derivatives of σ with respect to τ and λ (as required by (15) in §6, and the beginning of §7 of Chapter 2) can be summarized as follows. Each $\frac{\partial}{\partial \tau}$ derivative gains c/ρ_o, and each $\frac{\partial}{\partial \tau}$ derivative gains $c\{|t|/\rho_o + \frac{1}{\rho_o^2}\}$. Thus we want

$$(9) \qquad |\frac{\partial \sigma}{\partial \tau}| \leq c\rho_o^{-2}$$

$$(10) \qquad |\frac{\partial \sigma}{\partial \lambda}| \leq c\{|t|\rho_o^{-2} + \rho_o^{-3}\}$$

The reader should have no difficulty verifying (9) and (10) (and the corresponding inequalities for higher τ and λ derivatives) in the same way we proved (8). The differentiation condition with respect to t (see (17) in §7) requires that

$$\frac{\partial\sigma}{\partial t} = \lambda a_{-2} + \tau a'_{-2} + a_{-1} ,$$

where a_{-1} satisfies estimates like σ and a_{-2} and a'_{-2} satisfying similar estimates, but improved by a factor $c\rho_o^{-1}$ (i.e., $a_{-1} \in S^{-1}_{\rho_o}$, and a_{-2} and $a'_{-2} \in S^{-2}_{\rho_o}$). In fact the terms a'_{-2} and a_{-1} don't appear, and the fact that $|\frac{1}{\lambda}\frac{\partial}{\partial t}\sigma| \leq c\rho_o^{-2}$ can be easily verified using Lemmas 1 and 2. This, and the corresponding results for higher derivatives complete the proof that $\sigma \in S^{-1}_{\rho_o}$.

We are now in a position to write down the right parametrix for $\Omega_+ = i\xi_n + q(x, \xi)x_n$. (Recall that $q(x, \xi) = \left(\sum_{1 \leq i, j \leq n} a_{ij}(x)\xi_i\xi_j\right)^{1/2}$, where $\{a_{ij}(x)\}$ is a real symmetric positive definite matrix varying smoothly in x.) The parametrix is $P_+ = P$, where

(11) $\qquad P(x, \xi) = \sigma(x_n, \xi_n, q(x, \xi))$, with σ given by (5) and (6).

The ρ class associated with P is defined as follows: $Q(x, \xi)$ is the quadratic form $\xi_n^2 + x_n^2(\Sigma a_{ij}(x)\xi_i\xi_j)$ and $\rho^4 = (Q(x, \xi))^2 + \Sigma\xi_j^2 = Q^2 + |\xi|^2$.

Now unfortunately P is not of the class S^{-1}_{ρ}; this is not surprising, because the P is the presumptive inverse of $i\xi_n + x_n q(x, \xi)$ and this is not of the class S^1_ρ, since $x_n q(x,\xi) = x_n(\Sigma a_{ij}(x)\xi_i\xi_j)^{1/2}$ itself is not of the class S^1_ρ. What is needed is an extension of the classes S_ρ, which we now define.

Definition. $a(x,\xi)$ belongs to the "extended class S^m_ρ" $= (ES^m_\rho)$, if for every integer N we can write

$$a(x,\xi) = \sum_{j=1}^{n_N} a_j(x,\xi)\, b_j(x,\xi) + r_N(x,\xi) \tag{12}$$

where (a) $a_j(x,\xi) \in S^m_\rho$

(b) $b_j(x,\xi) \in S^0_{1,0}$

(c) $r_N(x,\xi)$ belongs to S^m_ρ when tested in terms of derivatives (in x and ξ) of order $\leq N$.

Remark. Since all the estimates we have made for operators of our symbol classes depend only on a finite number of derivatives, the required regularity properties will hold if in each case we make N large enough. Thus e.g., the operator whose symbol is $r_N(x,\xi)$ will map L^p_k to $L^p_{k-m/2}$, if N is sufficiently large (in terms of p, k and m).

Proposition 5. If $a \in ES^m_\rho$, then for each N we can write

$$a(x,\xi) = \sum_{j=1}^{n'_N} a'_j(x,\xi) \circ b'_j(x,\xi) + r'_N(x,\xi) \tag{12'}$$

where $a'_j(x,\xi) \in S^m_\rho$, $b'_j(x,\xi) \in S^0_{1,0}$, and $r'_N(x,\xi)$ belongs to S^m_ρ when tested in terms of derivatives of order $\leq N$.

It suffices to prove the proposition in the case $a(x,\xi) = a_1(x,\xi)\, b_1(x,\xi)$, with $a_1 \in S^m_\rho$, $b_1 \in S^0_{1,0}$. Now if $a_1 \in S^m_\rho$ then clearly $a_1 \in S^m_{1/2,1/2}$, and the symbolic calculus holds for $a_1 \circ b_1$. The Kohn-Nirenberg formula (see Hörmander [24], p. 147) states that

$$a_1 \circ b_1 = ab + \sum_{1 \le |\alpha| < M} \frac{\partial^\alpha_\xi a_1(x,\xi)\, \partial^\alpha_x b_1(x,\xi)}{i^{|\alpha|}\alpha!} + \text{remainder}$$

where the remainder belongs to $S^{m-M/2}_{1/2,1/2}$. Observe that each $\partial^\alpha_\xi a_1(x,\xi) \in S^{m-1}_\rho$, if $|\alpha| > 0$ and $\partial^\alpha_x b_1(x,\xi) \in S^0_{1,0}$, so we can iterate this process, until we finally obtain (12′), with $r'_N \in S^{m-M/2}_{1/2,1/2}$. If we choose M large enough we then see that $r'_N \in S^m_\rho$ as far as derivatives of order $\le N$ are concerned.

Our main result for the oblique derivative is as follows:

Theorem. (a) The symbol $P(x,\xi) = \sigma(x_n, \xi_n, q(x,\xi))$ (with σ given by (5) and (6)) belongs to ES^{-1}_ρ, and

(b) P is a right-parametrix for Ω_+, i.e., $\Omega_+ \circ P - 1 \in ES^{-1}_\rho$, (when restricted to a compact subset of the x-space).*

To prove the theorem we shall need a series of lemmas.

Lemma 3. Let φ be a C^∞ function on $(0,\infty)$ so that $\varphi(u) = 1$ when $0 \le u \le 2$, and $\varphi(u) = 0$, when $u \ge 5$. Then $\varphi(\rho_0^4/\lambda^2) \in S^0_{\rho_0}$.

---

*This restriction is necessary because we have always assumed all our symbols to have compact x support.

Recall that $\rho_o^4 = Q_0^2 + \lambda^2$, where $Q_0 = \tau^2 + t^2\lambda^2$. The condition on φ is such that $\varphi = 1$ when $Q_0 \leq \lambda$, and $\varphi = 0$ when $Q_0 \geq 2\lambda$.

The proof of the lemma follows easily from the fact that $\rho_o^m \in S^m_{\rho_o}$ (see the Appendix).

<u>Lemma 4.</u> <u>Suppose</u> $a(x,\xi) \in S^m_{1,0}$. <u>Then on the set where</u> $Q(x,\xi) \leq c\,|\xi|$, $a(x,\xi)$ <u>satisfies all the differential conditions of</u> S^{2m}_ρ.

<u>Proof.</u> Observe that $|\xi| \leq \rho^2 \equiv \{(Q(x,\xi))^2 + |\xi|^2\}^{1/2}$. Thus $|a(x,\xi| \leq c\rho^{2m}$ for large ξ. Moreover, each derivative with respect to ξ gives a gain of $O(|\xi|^{-1}) = O(\rho^{-2})$, since $Q(x,\xi) \leq c\,|\xi|$. So we have proved that if $a \in S^m_{1,0}$, then $a \in S^{2m}_\rho$ on the set where $Q \leq c\,|\xi|$. Now the x derivatives of a can be handled as follows. Write $1 = \sum_{j=1}^n \xi_j b_j(\xi) + b_0(\xi)$ where $b_j \in S^{-1}_{1,0}$, $j=1,\ldots,n$, and $b_0 \in S^0_{1,0}$. Then

$$\frac{\partial}{\partial x_\ell} a(x,\xi) = \sum_{j=1}^n \xi_j \left(\frac{\partial}{\partial x_\ell} a(x,\xi)\right) b_j(\xi) + \left(\frac{\partial a(x,\xi)}{\partial x_\ell}\right) b_0(\xi)$$

and by what we have just proved $\left(\frac{\partial a(x,\xi)}{\partial x_\ell}\right) b_j(\xi) \in \hat{S}^{2m-2}_\rho$ and $b_0(\xi)\frac{\partial a(x,\xi)}{\partial x_\ell} \in \hat{S}^{2m}_\rho$ on the set where $Q \leq c\,|\xi|$. The same argument works for higher x-derivatives. qed.

<u>Lemma 5.</u> <u>If</u> $a(t,\tau,\lambda) \in S^m_{\rho_0}$, <u>then</u> $a(x_n, \xi_n, a(x,\xi))$ <u>satisfies the condition of</u> S^m_ρ <u>on the set where</u> $Q \leq c\,|\xi|$.

Recall that $q(x,\xi) = (\sum a_{ij}(x)\xi_i\xi_j)^{1/2}$ where $\{a_{ij}(x)\}$ is real symmetric positive definite.

Proof. We start with a symbol $a(t, \tau, \lambda) \in S^m_{\rho_o}$ and substitute $\tau = \xi_n, t = x_n$, and $\lambda = q(x, \xi)$. Let us first verify that the resulting symbol is of class $\hat{S}^m_\rho$ in the set where $Q \leq c\, |\xi|$. Observe that $\rho_o(x_n, \xi_n, q(x, \xi))$ $= \{(\xi_n^2 + x_n^2 (q(x,\xi))^2)^2 + (q(x,\xi))^2\}^{1/2} = \rho(x, \xi)$, so $a(x_n, \xi_n, q(x, \xi))$ satisfies the correct bound. Next consider the ξ derivative. Taking $\frac{\partial}{\partial \xi_n}$ gives two terms, namely $(\frac{\partial}{\partial \tau} a)(x_n, \xi_n, q(x, \xi))$ and $(\frac{\partial}{\partial \lambda} a)(x_n, \xi_n, q(x,\xi)) \times \frac{\partial}{\partial \xi_n} q(x, \xi)$. But in both cases the decrease is (at least) $c/\rho_o \approx c/\rho$, so that condition is satisfied. If we form $\frac{\partial}{\partial \xi_j}$, $j \neq n$, then we get $(\frac{\partial}{\partial \lambda} a)(x_n, \xi_n, q(x, \xi)) \times \frac{\partial}{\partial \xi_j} q(x, \xi)$, and again the decrease is of the right order, this time $O\left(\frac{|x_n|}{\rho} + \frac{1}{\rho^2}\right)$. Higher ξ derivatives are treated in the same way, since we are dealing with products of symbols of the same kind with factors like $\frac{\partial}{\partial \xi_j} q(x, \xi)$. The latter is of course in $S^0_{1,0}$ and so behaves like an S^0_ρ symbol in the set where $Q \leq c\, |\xi|$, and so we see that $a(x_n, \xi_n, q(x, \xi)) \in \hat{S}^m_\rho$. Derivatives with respect to x are treated the same way. For example $\frac{\partial}{\partial x_n} a = (\frac{\partial}{\partial t} a)(x_n, \xi_n, q(x, \xi)) +$ $+ (\frac{\partial}{\partial \lambda} a)(x_n, \xi_n, q(x, \xi)) \frac{\partial q(x, \xi)}{\partial x_n}$. So

$$\frac{\partial}{\partial x_n} a = qb + \xi_n b' + b'' + \left(c \frac{x_n}{\rho} + \frac{c}{\rho^2}\right) \frac{\partial}{\partial x_n} q(x, \xi)$$

where $b, b' \in \hat{S}^{m-1}_\rho$, and $b'', c, c' \in \hat{S}^m_\rho$. We can write $q = \sum q_j \xi_j + q_0$, where $q_j \in S^0_{1,0}$ and $\frac{\partial q}{\partial x_n} = \sum \frac{\partial q_j}{\partial x_n} \xi_j + \frac{\partial q_0}{\partial x_n}$. Moreover $\rho^{-k} \in S^{-k}_\rho$ and

$x_n \xi_j \in S^1_\rho$. So $\frac{\partial}{\partial x_n} a$ satisfies the condition (17) in §7. The same argument works for (higher) x derivatives and the lemma is proved.

We now come to the proof of part (a) of the theorem. We write

$$P(x,\xi) = \varphi\left(\frac{\rho^4}{q^2}\right) P + \left(1 - \varphi\left(\frac{\rho^4}{q^2}\right)\right) P = P_1 + P_2 .$$

According to Lemmas 3 and 5, $\varphi\left(\frac{\rho^4}{q^2}\right) \in S^0_\rho$ and so also $1 - \varphi\left(\frac{\rho^4}{q^2}\right) \in S^0_\rho$. However, P_1 is supported in the set where $Q \leq 2q$, and so by lemmas, $P_1 \in S^{-1}_\rho$, since $\sigma \in S^{-1}_{\rho_o}$. We next decompose P_2 further, as $P_2 = P_3 + P_4$ where P_3 come from the term $i\sqrt{2} \exp\left\{-\frac{1}{2}\left(\frac{\tau^2 + t^2\lambda^2}{\lambda}\right)\right\} \lambda^{-1/2} \times e^{it\tau} \int_0^\infty e^{-s^2} e^{2\tau s/\sqrt{\lambda}} ds$, and P_4 from the term $-i \int_0^\infty e^{-\lambda s^2/2} e^{(\tau + it\lambda)s} ds$, (on making the substitution $t = x_n$, $\tau = \xi_n$, $\lambda = q(x,\xi)$).

We assert that $P_3 \in S^{-1}_\rho$, because derivatives with respect to ξ contribute either terms controlled by the $S^m_{\rho_o}$ calculus and lead to the correct decrease since $\rho_o \approx \rho$; or terms which arise when the ξ derivatives act on $q(x,\xi)$ or its derivatives. For each such derivative we get a gain of order $O(\lambda^{-1}) = O(q^{-1})$, and so for a derivative of order m we are led to estimating $\lambda^{-m-1/2} \exp\left\{-\frac{1}{2}\left(\frac{\tau^2 + t^2\lambda^2}{\lambda}\right)\right\}$; however, this is clearly $O(\tau^2 + t^2\lambda^2)^{-m-1/2} = O(\rho^{-2m-1})$, since we are restricted to the set where $\tau^2 + t^2\lambda^2 \geq \lambda$, i.e., where $\tau^2 + t^2\lambda^2 \approx \rho^2$. Similar arguments hold for the x derivatives.

Finally we come to the term P_4 which involves consideration of the integral

$$\lambda^{-1/2}\int_0^\infty e^{-s^2/2}e^{zs}\,ds, \qquad \text{where } z = \frac{\tau + it\lambda}{\sqrt{\lambda}}$$

and $\mathrm{Re}(z) \le 0$. We claim the following asymptotic expansion

$$(13) \qquad \int_0^\infty e^{-s^2/2}e^{zs}\,ds \sim \sum_{k=0}^{\infty} c_k z^{-2k-1}, \quad \text{with } c_k \text{ real,}$$

in the sense that $\left(\frac{d}{dz}\right)^{\ell}\left[\int_0^\infty e^{-s^2/2}e^{zs}\,ds - \sum_{0\le k\le m} c_k z^{-2k-1}\right] = O(|z|^{-2m-\ell-1})$

as $z \longrightarrow \infty$, $\mathrm{Re}(z) \le 0$. (13) can be proved by writing

$$\int_0^\infty e^{-s^2/2}e^{zs}\,ds = \int_0^\infty \psi(s)\,e^{-s^2/2}e^{zs}\,ds + \int_0^\infty (1-\psi(s))\,e^{-s^2/2}e^{zs}\,ds$$

where ψ is a C^∞ function which is $=1$ near zero, and vanishes outside a compact set. The second integral contributes zero asymptotically (see the proof of Lemma 1), while if we use the fact that

$e^{-s^2/2} = \sum(-1)^k \frac{2^{-k}}{k!}s^{2k}$, we get after integration by parts (13) with

$c_k = (-1)^{k+1}\frac{(2k)!}{2^k k!}$. This allows us to write

$$iP_4 = (1-\varphi(\rho^4/q))\sum_{k=0}^{m} c_k\left(\frac{\tau+it\lambda}{\lambda^{1/2}}\right)^{-(2k+1)}\lambda^{-1/2} + \text{remainder} = iP_5 + iP_6.$$

The series can be written as $\sum_{k=0}^{m} c_k \frac{(\tau - it\lambda)^{2k+1}\lambda^k}{(\tau^2+t^2\lambda^2)^{2k+1}}$. Collecting terms

we get linear combinations of terms of the form $\frac{\tau^a(t\lambda)^b\lambda^k}{(\tau^2+t^2\lambda^2)^{2k+1}}$.

When we substitute $\lambda = q(x,\xi)$, we observe that we can write it in the form $\sum_{j=1}^{n} \xi_j q_j(x,\xi) + q_0$, where $q_0, q_j \in S^0_{1,0}$. Moreover $\tau = \xi_n$, and $x_n \xi_j$ are symbols in S^1_ρ, while ξ_j, $j=1,\ldots,n-1$ belong to S^2_ρ. In addition, restricted to the set where $Q \geq c\lambda$ (i.e., $\tau^2 + t^2\lambda^2 \geq c\lambda$) we have that $(\tau^2+t^2x^2)^{-1} = (\xi_n^2 + x_n^2 q^2(x,\xi))^{-1} \in S^{-2}_\rho$, as we have already pointed out. Altogether then except for the remainder term, iP_4 is a finite sum of products symbols of the class S^{-1}_ρ and $S^0_{1,0}$.

Now the remainder term insofar as it and its derivatives are concerned behaves no worse than $\dfrac{\lambda^{m+1/2}}{(\tau+it\lambda)^{2m+1}}$, which behaves like a symbol of class S^{-1}_ρ when derivatives of order $\leq m$ are considered, (with of course $\tau = \xi_n$, $t = x_n$, $\lambda = q(x,\xi)$). This completes the proof of the fact that $P \in ES^{-1}_\rho$.

(The reader should keep in mind that in view of the symmetry property (6) for σ, it suffices to carry out our computations for P when $\tau = \xi_n \leq 0$ as long as the decompositions we have carried out respects this symmetry. Notice that this is the case for P_1 (and P_2) by their definition. Now $P_4 = P_5 + P_6$, and P_5 has this symmetry, because coefficients c_k are real and the exponents $2k+1$ are odd. Since $P = P_1 + P_2$, $P_2 = P_3 + P_4$, we have $P = P_1 + P_3 + P_5 + P_6$, so $P - P_1 - P_5 = P_3 + P_6$ and this is extended by the symmetry (6) to $\tau \geq 0$.)

We shall compute $\Omega_+ \circ P$ in order to prove part (b) of the theorem. We have observed that $\Omega_+ \in S^1_{1,0}$, and since $P \in ES^{-1}_\rho$, we have that P also belongs to $S^{-1/2}_{1/2,1/2}$. Thus we can apply the symbolic calculus and obtain (see (1))

$$(14)\qquad \Omega_+ \circ P = (i\xi_n + q(x,\xi)_n)P(x,\xi) + \frac{\partial P}{\partial x_n}(x,\xi) + R_1 + R_2$$

where R_1 is a sum of terms involving products of the form

$$(15)\qquad \left(\frac{\partial}{\partial \xi}\right)^{\alpha}[q(x,\xi)\,x_n]\left(\frac{\partial^{\alpha}}{\partial x}\right)P(x,\xi),\quad \text{with } |\alpha| > 0 \text{ and } R_2 \in S^{-m}_{1/2,1/2},$$

with preassigned m.

However, $\sigma(t,\tau,\lambda)$ is the symbol of a right inverse of $\frac{d}{dt} + t\lambda$, thus we have

$$(16)\qquad (i\tau + \lambda t)\sigma(t,\tau,\lambda) + \frac{\partial \sigma}{\partial t}(t,\tau,\lambda) \equiv 1$$

Substituting $\tau = \xi_n$, $t = x_n$, and $\lambda = q(x,\xi)$ gives us

$$\Omega_+ \circ P = 1 + \frac{\partial \sigma}{\partial \lambda}(x_n, \xi_n, q(x,\xi))\frac{\partial}{\partial x_n}(q(x,\xi)) + R_1 + R_2.$$

Now from the formula (5) it is evident that $\frac{\partial \sigma}{\partial \lambda}$ is of the form $t\sigma' + \sigma''$ where σ' and σ'' have similar expressions to σ but $\sigma' \in S^{-2}_{\rho_0}$, and $\sigma'' \in S^{-3}_{\rho_0}$. Thus $\sigma'(x_n, \xi_n, q(x,\xi)) \in ES^{-2}_{\rho}$ and $\sigma''(x_n, \xi_n, q(x,\xi)) \in ES^{-3}_{\rho}$, by the same argument used to treat $P(x,\xi) = \sigma(x_n, \xi_n, q(x,\xi))$. Moreover, using the fact that $x_n\xi_j \in S^1_\rho$, $\xi_j \in S^2_\rho$ and $q(x,\xi) = \sum q_j(x,\xi)\xi_j + q_0(x,\xi)$, where $q_0, q_j \in S^0_{1,0}$ we see that $\frac{\partial \sigma}{\partial \lambda}(x_n, \xi_n, q(x,\xi))\frac{\partial}{\partial x_n}q(x,\xi) \in ES^{-1}_{\rho}$. By the same argument the terms (15) also belong to ES^{-1}_{ρ}. Finally, if we take m sufficiently large, then every symbol in $S^{-m}_{1/2,1/2}$ belongs to S^{-1}_{ρ} when tested against derivatives of order $\leq N$. This shows that

$\Omega_+ P - 1 \in ES_\rho^{-1}$ and the theorem is proved.

Concluding remarks

(1). We can obtain a similar left-parametrix for $\Omega_- = i\xi_n - q(x,\xi)x_m$. This can be done by taking adjoints of the basic relation $\Omega_+ \circ P - 1 \in ES_\rho^{-1}$ and noting that all the symbol classes used are essentially invariant under adjoints. Alternatively we can follow a parallel derivation to that for Ω_+ by first computing the symbol $\widetilde{\sigma}(t,\tau,\lambda)$ of the operator $\widetilde{K}_\lambda$ which satisfies $\widetilde{K}_\lambda(\frac{d}{dt} - \lambda t) = I$ $(\lambda > 0)$, and which is uniquely specified by the further fact that $\widetilde{K}_\lambda(e^{-\lambda t^2/2}) = 0$. One can show (with an argument similar to that of Proposition 3), that

$$(5') \qquad \widetilde{\sigma}(t,\tau,\lambda) = \sqrt{2}\, e^{-\frac{1}{2}\left[\frac{\tau^2+\lambda^2 t^2}{\lambda}\right]} e^{it\lambda} \int_0^\infty e^{-\lambda s^2} e^{-2\lambda t s}\, ds - \int_0^\infty e^{-\lambda s^2/2} e^{-s(\lambda t + i\tau)}\, ds, \qquad t \geq 0$$

$$(6') \qquad \widetilde{\sigma}(-t,\tau,\lambda) = -\overline{\widetilde{\sigma}(t,\tau,\lambda)}$$

(2). The assumption A_1 for the symbols S_ρ^m (as given in §9) holds in this case. In fact if we take $\mathcal{L} = \frac{\partial^2}{\partial x_n^2} + x_n^2 \sum_{j=1}^n \frac{\partial^2}{\partial x_j^2}$, then according to Theorem 21 in §14 this operator has a left parametrix E in S_ρ^{-2}, and we need only take $\{P_j(x,D)\} = \{\frac{\partial}{\partial x_n}, x_n \frac{\partial}{\partial x_j}, 1 \leq j \leq n\}$ while $Q_j(x,D) = E \cdot P_j(x,D)$. It is also to be noted that the non-isotropic Sobolev spaces

Sob^p_k discussed here are then equivalent to the spaces S^p_k for the operator $\mathcal{L}$, treated in Rothschild-Stein [39].

(3). As a result of the above, and in particular the first remark, we can assert the following local regularity result for the operator Ω_- (the operator Ω_+ has a corresponding existence statement). Suppose

(17) $$\Omega_- f = g$$

and g belongs (locally) to either L^p_k, Λ_α, or Sob^p_k. Then f belongs (locally) to $L^p_{k+1/2}$, $\Lambda_{\alpha+1/2}$, or Sob^p_{k+1}, respectively. This follows from the regularity results of the class S^{-1}_ρ, and the fact that operators whose symbols belong to $S^0_{1,0}$ preserve these classes. However, we do not know if $g \in \Gamma_\alpha$ implies that $f \in \Gamma_{\alpha+1}$, because it is not true that operators of the standard class $S^0_{1,0}$ preserve Γ_α.

§16. Second-order operators of Kannai-type

We shall construct the parametricies for operators $L_\pm$ of the form

(1) $$L_\pm = \frac{\partial}{\partial x_0} + \sum_{j=1}^{n} a_j(x)\frac{\partial}{\partial x_j} \pm x_0 \sum_{1 \le i,j \le n} a_{ij}(x)\frac{\partial^2}{\partial x_i \partial x_j}$$

where $a_i(x)$ and $a_{ij}(x)$ are smooth real functions, with the $n \times n$ matrix $\{a_{ij}(x)\}$ symmetric and positive definite.

Kannai [26] showed that basic examples of the operators L_+ of the type (1) are unsolvable, yet hypoelliptic. This result has since been extended by several people, see e.g., Beals and C. Fefferman [4] where some earlier references may be found.

We shall show how to construct a right parametrix for L_- (and similarly a left parametrix for L_+). The method will be similar to that used for the oblique derivative problem in the previous section, but the details will turn out to be much simpler. Let us deal first with L_-. We begin by describing the symbol classes appropriate for this problem.

We take $Q^1(x,\xi)$ to be $\sum_{1\le i,j\le n} a_{ij}(x)\,\xi_i\xi_j$ which we can write as $\sum_{i=1}^{n} |L_i(x)\xi|^2$, where $\xi \longrightarrow L_i(x)\xi$ are a spanning set of linear forms for the subspace given by $\xi_0 = 0$, and which depend smoothly on x. The function $\varphi(x)$ equals x_0, and we set $Q(x,\xi) = \varphi(x)\,Q^1(x,\xi)$. The ρ function is defined by $\rho(x,\xi) = \{(Q(x,\xi))^2 + |\xi|^2\}^{1/4}$, hence

(2) $$\rho(x,\xi) \approx |x_0|^{1/2}\,|\xi'| + |\xi|^{1/2}, \quad \text{where } |\xi'|^2 = \sum_{j=1}^{n} \xi_j^2$$

The natural candidate for the symbol $P(x,\xi)$ of a right parametrix of L_- is

(3) $$P(x,\xi) = \sigma(x_0,\, \xi_0 + \sum_{j=1}^{n} a_j(x)\xi_j,\, \sum_{1\le j,k\le n} a_{jk}(x)\,\xi_j\xi_k),$$

with σ given by (5) and (6) of §15.

Theorem. When restricted to compact x subsets

(a) $P(x,\xi) \in S_\rho^{-2}$

(b) $$\Big(i(\xi_0 + \sum_{j=1}^{n} a_j(x)\xi_j) + x_0 \sum_{1\le j,k\le n} a_{jk}(x)\xi_j\xi_k\Big) \circ P(x,\xi) - 1 \in S_\rho^{-1}$$

Proof. Recall that $\rho_0 = \{(\tau^2 + t^2\lambda^2)^2 + \tau^2 + \lambda^2\}^{1/4}$, and if we make the substitution $\tau = \xi_0 + \Sigma a_j(x)\xi_j$, $t = x_0$, $\lambda = \sum_{1 \le j,k \le n} a_{jk}(x)\xi_j\xi_k$, we see that

(4) $$\rho_o \approx \rho^2$$

Thus by (8) of §15, we see that

(5) $$|P(x,\xi)| \le c\rho^{-2}$$

Let us now consider the ξ-derivatives of P. The requirement that P belongs to S_ρ^{-2} is that an application of $\frac{\partial}{\partial \xi_0}$ gives a gain of $O(1/\rho^2)$, while an application of $\frac{\partial}{\partial \xi_j}$, j=1,...,n, should give a gain of $\frac{|x_0|^{1/2}}{\rho} + \frac{1}{\rho^2}$. Thus we expect

(6) $$\left|\frac{\partial}{\partial \xi_0} P(x,\xi)\right| \le c\,\rho^{-4}$$

and this follows from (9) in §15, since $\rho^2 \approx \rho_o$. We also expect

(7) $$\left|\frac{\partial}{\partial \xi_j} P(x,\xi)\right| \le c\left(\frac{|x_0|^{1/2}}{\rho^3} + \frac{1}{\rho^4}\right), \quad j=1,\ldots,n.$$

However, $\frac{\partial}{\partial \xi_j} P(x,\xi) = \frac{\partial\sigma}{\partial\tau}\cdot a_j(x) + \frac{\partial\sigma}{\partial\lambda}\frac{\partial}{\partial \xi_j}(Q^1(x,\xi))$. The term $\frac{\partial\sigma}{\partial\tau}a_j(x)$ is $O(1/\rho^4)$, because of (9) in §15 and $\rho^2 \approx \rho_o$. While $\frac{\partial}{\partial \xi_j}(Q^1(x,\xi)) \le c\,(Q^1(x,\xi))^{1/2}$, and $\frac{\partial\sigma}{\partial\lambda}$ is bounded by $c\left\{\frac{|x_0|}{\rho^4} + \frac{1}{\rho^6}\right\}$. Now $|x_0|^{1/2}Q^1(x,\xi)^{1/2} \le P$ while $\left|\frac{\partial}{\partial \xi_j}Q^1(x,\xi\right| \le c\;|\xi| \le c\rho^2$. This gives us (7).

The higher ξ derivatives are treated in the same way. Let us now consider the x derivatives. We claim that

(8) $$\frac{\partial}{\partial x_0} P(x, \xi) = Q^1(x, \xi)a + \sum_{j=1}^{n} b_j \xi_j + c$$

where $a = O(\rho^{-4})$, $b_j \in O(\rho^{-3})$, and $c = O(\rho^{-2})$. (More precisely $a \in \hat{S}_{\rho}^{-4}$, $b_j \in \hat{S}_{\rho}^{-3}$, and $c \in \hat{S}_{\rho}^{-2}$.) In fact, $\frac{\partial}{\partial \lambda_0} P(x, \xi) = \frac{\partial \sigma}{\partial t} + \frac{\partial \sigma}{\partial \tau}(\sum_j \frac{\partial}{\partial x_0}(a_j(x)\,\xi_j) + \frac{\partial \sigma}{\partial \lambda} \frac{\partial}{\partial x_0} Q^1(x, \xi)$, and so (8) follows from what we have proved in §15 for $\frac{\partial \sigma}{\partial t}$, $\frac{\partial \sigma}{\partial \tau}$, and $\frac{\partial \sigma}{\partial \lambda}$, together with the observations that

$$|x_0 \frac{\partial Q^1}{\partial x_0}(x, \xi)| \leq c\,|x_0 \sum_{j=1}^{n} \xi_j^2| \leq c\,\rho^2, \quad \text{while}$$

$$|\frac{\partial Q^1}{\partial x_0}(x, \xi)| \leq c\,\rho^4.$$

Of course (8) is of the appropriate form as required by definition (20) in §7.

By the same argument we see that

(9) $$\frac{\partial}{\partial x_k} P(x, \xi) = \sum_{j=1}^{n} b_j' \xi_j + c', \qquad k=1, \ldots, n,$$

where again $b_j' = O(\rho^{-3})$, $c' = O(\rho^{-2})$, since here the term $\frac{\partial \sigma}{\partial t}$ does not enter. Higher x derivatives are treated similarly. Thus, finally, $P \in S_{\rho}^{-2}$

To prove part (b) of the theorem we use the identity (16) of §15 and the symbolic calculus. Thus

$$\left(i(\xi_0 + \sum_{j=1}^{n} a_j(x)\xi_j) + x_0 \sum_{1 \leq j, k \leq n} a_{jk}(x)\,\xi_j \xi_k\right) \circ P(x, \xi) - 1$$

is a sum of terms of the form

(10) $$\frac{\partial P}{\partial x_j}, \quad x_0\xi_k \frac{\partial P}{\partial x_j}, \quad \text{and} \quad x_0 \frac{\partial^2 P}{\partial x_j \partial x_k}, \qquad 1 \leq j \leq k \leq n$$

multiplied by smooth functions of x. By (9), and since actually $b'_j \in S_\rho^{-3}$ and $c' \in S_\rho^{-2}$, we see that $\frac{\partial P}{\partial x_j} \in S_\rho^{-1}$ and $x_0\xi_k \frac{\partial P}{\partial x_j} \in S_\rho^{-1}$. (Recall that $\xi_j \in S_\rho^2$ while $x_0\xi_k\xi_j \in S_\rho^2$.) The analogue of (9) for second-derivatives (excluding derivation with respect to x_0) is

$$\frac{\partial^2 P}{\partial x_j \partial x_n} = \sum_{1 \leq j, \ell \leq n} b_{i\ell}\xi_j\xi_\ell + \sum_{1 \leq \ell \leq n} c_\ell \xi_\ell + d$$

where $b_{i\ell} \in S_\rho^{-4}$, $c_\ell \in S_\rho^{-3}$ and $d \in S_\rho^{-2}$.

However, $x_0\xi_j\xi_\ell \in S_\rho^2$ and $\xi_\ell \in S_\rho^2$, so all the terms which appear in (10) are in S_ρ^{-1}, proving the theorem.

Further remarks

(1) Suppose $u(t,x)$ is a solution of the heat equation $\frac{\partial u}{\partial t} - \sum_{j=1}^{n} \frac{\partial^2}{\partial x^2} u(t,x) = 0$, with $u \in C^2$ $(t \geq 0,\ x \in \mathbb{R}^n)$. Then

(11) $$f(x_0, x) = u(x_0^2/2,\ x)$$

is a solution of $L_-(f) = 0$, in the case $a_i \equiv 0$, $a_{jk} = \delta_{jk}$. Moreover, if $u(0,x) \notin C^\infty$, then $f \notin C^\infty$, which shows that this L_- is not hypoelliptic. In this special case of L_-, $P(x,\xi) = \sigma(x_0, \xi_0, \sum_{j=1}^{n} \xi_j^2)$ is the exact symbol of the operator which is the right inverse of L_- uniquely determined by $L_- P(x, \frac{\partial}{i\partial x}) = I$, and the range of $P(x, \frac{\partial}{i\partial x})$ is orthogonal to the null solutions of L_- of the form (11).

(2) By similar methods (using $\tilde{\sigma}$ instead of σ) we can obtain a left parametrix for L_+, whose symbol also belongs to S_ρ^{-2}. Again, in the special case when $a_i \equiv 0$, $a_{jk} = \delta_{jk}$, then $\tilde{P}(x,\xi) = \tilde{\sigma}(x_0, \xi_0, \sum_{j=1}^{n} \xi_j^2)$ is the exact symbol of the operator which is the left inverse of L_+ uniquely determined by $\tilde{P}(x, \frac{\partial}{i\partial x}) L_+ = I$, and $P(x, \frac{\partial}{i\partial x})$ annihilates the functions of the form (11).

(3) The parametrix for the general L_+ proves the following regularity theorem. Suppose $L_+ f = g$, and g belongs (locally) to L_k^p, Λ_α, Sob_k^p, or Γ_α; then f belongs locally to L_{k+1}^p, $\Lambda_{\alpha+1}$, Sob_{k+2}^p, or $\Gamma_{\alpha+2}$, respectively.

Appendix: Some computations concerning the class S^m_ρ

In order not to have discouraged the interested reader we postponed to this appendix some of the more elementary but tiresome computations involving our symbols. Recall the definitions of the symbol class S^m_ρ and the preliminary classes $\hat{S}^m_\rho$, $\tilde{S}^m_\rho$ given in §6 and §7. Among these hold the inclusion relations $\hat{S}^m_\rho \supset \tilde{S}^m_\rho \supset S^m_\rho$, (of which the first is obvious, and the second is proved below).

§1. Preliminaries

Let $A(x)$ be a real symmetric $n \times n$ matrix, and suppose for each x, $A(x)$ is either positive or negative semi-definite. Put $Q(x, \xi) = (A(x)\xi, \xi)$. If we let $(\nabla_\xi, \eta) = \sum \eta_j \frac{\partial}{\partial \xi_j}$, we then have

(a) $$(\nabla_\xi, \eta)\, Q(x, \xi) = 2\,(A(x)\,\xi, \eta)$$

(b) $$(\nabla_\xi, \eta_1)(\nabla_\xi, \eta_2)\, Q(x, \xi) = 2\,(A(x)\eta_1, \eta_2)$$

(c) $$\prod_{j=1}^{k} (\nabla_\xi, \eta_j)\, Q(x, \xi) = 0 \quad \text{if } k \geq 3$$

Since $A(x)$ is semi-definite we have $|(A(x)\xi, \eta)| \leq |(A\xi, \xi)|^{1/2}\, |(A\eta, \eta)|^{1/2}$. Hence:

(i) $$|(\nabla_\xi, \eta)\, Q(x, \xi)| \lesssim |Q(x, \xi)|^{1/2}\, |Q(x, \eta)|^{1/2}$$

(ii) $$|(\nabla_\xi, \eta_1)(\nabla_\xi, \eta_2)\, Q(x, \xi)| \lesssim |Q(x, \eta_1)|^{1/2}\, |Q(x, \eta_2)|^{1/2}$$

(iii) $$\prod_{j=1}^{k} (\nabla_\xi, \eta_j)\, Q(x, \xi) = 0 \quad \text{if } k \geq 3.$$

Next, we have already checked that if $Q(x, \xi) = \varphi(x)\, Q^1(x, \xi)$ as before,

Then (Proposition 7)

(iv) $$|(\nabla_x, \eta)\, Q(x,\xi)| \lesssim |\xi|\rho(x,\xi)\, N(x,\eta) + |\xi|^2\, N(x,\xi)^2$$

Since Q is a quadratic form depending smoothly on x we also have:

(v) $$\Big|\prod_{j=1}^{k} (\nabla_x, \eta_j)\, Q(x,\xi)\Big| \lesssim |\xi|^2 \prod_{j=1}^{k} |\eta_j|$$

§2. The class $\tilde{S}^m_\rho$

Lemma A. Assume that $Q^1(x,\xi) = \sum_j [L_j(x,\xi)]^2$ where each $L_j(x,\xi)$ is a smoothly varying linear functional in ξ. Then for $\rho(x,\xi) \geq 1$,

$$\Big|\prod_{j=1}^{\ell} (\nabla_\xi, \eta_j) \prod_{k=1}^{m} (\nabla_x, \lambda_k)\, Q(x,\xi)\Big|$$

$$\lesssim \rho(x,\xi)^2 \prod_{j=1}^{\ell}\Big[\frac{\rho(x,\eta_j)}{\rho(x,\xi)} + \frac{\rho(x,\eta_j)^2}{\rho(x,\xi)^2}\Big] \prod_{k=1}^{m}\Big[\rho(x,\xi)\, N(x,\lambda_k) + \rho(x,\xi)^2 N(x,\lambda_k)^2\Big]$$

Proof. We consider possible values for ℓ and m:

$\ell = 0,\ m = 0$ follows since $(Q(x,\xi)) \lesssim \rho(x,\xi)^2$.

$\ell = 0,\ m = 1$ follows from (iv) since $\dfrac{|\xi|}{\rho(x,\xi)} \lesssim \rho(x,\xi)$.

$\ell = 0,\ m \geq 2$. We have by (v)

$$\Big|\prod_{k=1}^{m} (\nabla_x, \lambda_k)\, Q(x,\xi)\Big| \lesssim |\xi|^2 \prod_{k=1}^{m} (\lambda_k)$$

$$\lesssim \rho(x,\xi)^2 \Big[\frac{|\xi|}{\rho(x,\xi)}\Big]^2 \prod_{k=1}^{m} N(x,\lambda_k)$$

$$\lesssim \rho(x,\xi)^2 \prod_{k=1}^{m} [\rho(x,\xi)\, N(x,\lambda_k)]$$

since $m \geq 2$, and $\rho(x,\xi) \geq 1$. This completes the case $\ell = 0$.

$\ell = 1,\ m = 0$: By (i)

$$|(\nabla_\xi, \eta)\, Q(x,\xi)| \lesssim |Q(x,\xi)|^{1/2}\ |Q(x,\eta)|^{1/2}$$

$$\lesssim \rho(x,\xi)^2 \left[\frac{|Q(x,\xi)|^{1/2}}{\rho(x,\xi)^2}\ |Q(x,\eta)|^{1/2}\right]$$

$$\lesssim \rho(x,\xi)^2 \left[\frac{\rho(x,\eta)}{\rho(x,\xi)}\right]$$

We will skip the case $\ell = 1,\ m = 1$ until the end.

$\ell = 1,\ m \geq 2$:

$$|(\nabla_\xi, \eta) \prod_{k=1}^{m} (\nabla_x, \lambda_k)\, Q(x,\xi)|$$

$$= \left| \prod_{k=1}^{m} (\nabla_x, \lambda_k)\, (A(x)\,\xi, \eta) \right|$$

$$\lesssim \prod_{k=1}^{m} |\lambda_k|\ |\xi|\ |\eta|$$

$$\lesssim \rho(x,\xi)^2 \left[\frac{|\eta|}{\rho(x,\xi)^2}\right] \prod_{k=1}^{m} [N(x,\lambda_k)]\ \rho(x,\xi)^2$$

(since $|\xi| \lesssim \rho(x,\xi)^2$).

$$\lesssim \rho(x,\xi)^2 \left[\frac{\rho(x,\eta)^2}{\rho(x,\xi)^2}\right] \prod_{k=1}^{m} [\rho(x,\xi)\, N(x,\lambda_k)]$$

since $m \geq 2$.

$\ell = 2,\ m = 0$: By (ii)

$$|(\nabla_\xi, \eta_1)\,(\nabla_\xi, \eta_2)\, Q(x,\xi)| \lesssim Q(x,\eta_1)^{1/2}\ Q(x,\eta_1)^{1/2}$$

$$\lesssim \rho(x,\xi)^2 \left[\frac{\rho(x,\eta_1)}{\rho(x,\xi)}\right] \left[\frac{\rho(x,\eta_1)}{\rho(x,\xi)}\right.$$

$\ell = 2$, $m = 1$: We skip this also until the end.

$\ell = 2$, $m \geq 2$

$$|(\nabla_\xi, \eta_1)(\nabla_\xi, \eta_2) \prod_{k=1}^{m} (\nabla_x, \lambda_k) Q(x, \xi)| \lesssim |\eta_1| \, |\eta_2| \prod_{k=1}^{m} |\lambda_k|$$

$$\lesssim \rho(x,\xi)^2 \frac{|\eta_1|}{\rho(x,\xi)^2} \frac{|\eta_2|}{\rho(x,\xi)^2} \Big[\prod_{k=1}^{m} |\lambda_k|\Big] \rho(x,\xi)^2$$

$$\lesssim \rho(x,\xi) \Big[\frac{\rho(x,\eta_1)^2}{\rho(x,\xi)^2}\Big] \Big[\frac{\rho(x,\eta_2)^2}{\rho(x,\xi)^2}\Big] \prod_{k=1}^{m} [\rho(x,\xi) N(x,\lambda_k)]$$

since $m \geq 2$.

When $\ell > 2$, the derivative is identically zero, so there is nothing to prove. Hence it only remains to check the cases $m = 1$, $\ell = 1$ and $m = 1$, $\ell = 2$. Here we use $Q(x,\xi) = \varphi(x) Q^1(x,\xi)$ where Q^1 is a sum of squares.

$$\prod_j (\nabla_\xi, \eta_j)(\nabla_x, \lambda) Q(x,\xi)$$

$$= \prod_j (\nabla_\xi, \eta_j) [\varphi(x)(\nabla_x, \lambda) Q^1(x,\xi) + (\nabla\varphi(x), \lambda) Q^1(x,\xi)]$$

$$= \varphi(x) \prod_j (\nabla_\xi, \eta_j)(\nabla_x \lambda) Q^1(x,\xi) + (\nabla\varphi(x), \lambda) \prod_{j=1}^{m} (\nabla_\xi, \eta_j) Q^1(x,\xi)$$

We deal with the second term first. As in Proposition 7,

$$|(\nabla\varphi(x), \lambda)| \lesssim [N(x,\lambda)\,|\varphi(x)| + N(x,\lambda)^2] \qquad (\text{since } |\nabla\varphi(x)| \lesssim c)$$

When $m = 1$, we set

$$\lesssim [N(x,\lambda)\,|\varphi(x)| + N(x,\lambda)^2]\, Q^1(x,\xi)^{1/2} Q^1(x,\eta)^{1/2}$$

$$\lesssim N(x,\lambda)|Q(x,\xi)|^{1/2} \, |Q(x,\eta)|^{1/2} + N(x,\lambda)^2 \, |\xi| \, |\eta|$$

$$\lesssim \rho(x,\xi)^2\left[\frac{\rho(x,\eta)}{\rho(x,\xi)}\right]N(x,\lambda) + \rho^2(x,\xi)\left[\frac{\rho(x,\eta)^2}{\rho(x,\xi)^2}\right]\left[\rho(x,\xi)^2 N(x,\lambda)^2\right]$$

$$\lesssim \rho(x,\xi)^2\left[\frac{\rho(x,\eta)}{\rho(x,\xi)} + \frac{\rho(x,\eta)^2}{\rho(x,\xi)^2}\right]\left[\rho(x,\xi)\, N(x,\lambda) + \rho(x,\xi)^2 N(x,\lambda)^2\right].$$

When $m = 2$, we get

$$\lesssim [N(x,\lambda)\,|\varphi(x)| + N(x,\lambda)^2\, Q^1(x,\eta_1)^{1/2}\, Q^1(x,\eta_2)^{1/2}$$

$$\lesssim N(x,\lambda)|Q(x,\eta_1)|^{1/2}|Q(x,\eta_2)|^{1/2} + N(x,\lambda)^2\ |\eta_1|\ |\eta_2|$$

$$\lesssim \rho(x,\xi)^2\left[\frac{\rho(x,\eta_1)}{\rho(x,\xi)}\right]\left[\frac{\rho(x,\eta_2)}{\rho(x,\xi)}\right]N(x,\lambda) +$$

$$+ \rho(x,\xi)^2\left[\frac{\rho(x,\eta_1)^2}{\rho(x,\xi)^2}\right]\left[\frac{\rho(x,\eta_2)^2}{\rho(x,\xi)^2}\right]\left[N(x,\lambda)^2\rho(x,\xi)^2\right].$$

Hence we have to take care of $\varphi(x)\prod_{j=1}^{m}(\nabla_\xi,\eta_j)(\nabla_x\lambda)\,Q^1(x,\xi)$

where $Q^1 = \sum_j (L_j(x,\xi))^2$. Now for

$m = 1$, we get $\varphi(x)\sum(\nabla_\xi,\eta)(\nabla_x,\lambda)(L(x,\xi)^2)$

$$= 2\varphi(x)\sum[\ell(x,\xi)(\nabla_\xi,\eta)(\nabla_x\lambda)\,L(x,\xi)$$

$$+ (\nabla_\xi,\eta)\,L(x,\xi)\,(\nabla_x\lambda)\,L(x,\xi),$$

and this is bounded by $\varphi(x)\,[Q^1(x,\xi)^{1/2}\ |\eta|\ |\lambda| + Q^1(x,\eta)^{1/2}\ |\xi|\ |\lambda|]$

$$\lesssim Q(x,\xi)^{1/2}\,|\eta|\ |\lambda| + Q(x,\eta)^{1/2}\,|\xi|\ |\lambda|$$

$$\lesssim \rho(x,\xi)^2\left[\frac{\rho(x,\eta)^2}{\rho(x,\xi)^2}\right][N(x,\lambda)\,\rho(x,\xi)] +$$

$$+ \rho(x,\xi)^2\left[\frac{\rho(x,\eta)}{\rho(x,\xi)}\right]\left[N(x,\lambda)\,\rho(x,\xi)\right].$$

For $m = 2$, we get $\varphi(x)\sum_j (\nabla_\xi, \eta_1)(\nabla_\xi\ \eta_2)(\nabla_x\lambda)(L_j(x,\xi))^2$

$$= 2\varphi(x)\sum(\nabla_\xi\eta_1)(\nabla_\xi\eta_2)\,[L_j(x,\xi)(\nabla_x\lambda)L_j(x,\xi)]$$

$$= 2\varphi(x)\sum(L_j(x,\xi)(\nabla_\xi,\eta_1)(\nabla_\xi,\eta_2)(\nabla_x,\lambda)L_j$$

$$+ (\nabla_\xi\eta_1)L_j(x,\xi)(\nabla_\xi\eta_2)(\nabla_2\lambda)L_j(x,\xi)$$

$$+ (\nabla_\xi\eta_2)L_j(x,\xi)(\nabla_\xi\eta_1)(\nabla_x,\lambda)L_j(x,\xi)$$

$$+ (\nabla_\xi\eta_1)(\nabla_\xi\eta_2)L_j(x,\xi)(\nabla_x,\lambda)L_j),$$ which is bounded by

$$\lesssim \varphi(x)\,[Q^1(x,\eta_1)^{1/2}\,|\eta_2|\,|\lambda| + Q^1(x,\eta_2)^{1/2}\,|\eta_1|\,|\lambda|]$$

$$\leq |Q(x,\eta_1)|^{1/2}\,|\eta_2|\,|\lambda| + |Q(x,\eta_2)|^{1/2}\,|\eta_1|\,|\lambda|$$

$$\lesssim \rho(x,\xi)^2\left[\frac{\rho(x,\eta_1)}{\rho(x,\xi)}\right]\left[\frac{\rho(x,\eta_2)^2}{\rho(x,\xi)^2}\right][N(x,\lambda)\,\rho(x,\xi)]$$

$$+ \rho(x,\xi)^2\left[\frac{\rho(x,\eta_1)^2}{\rho(x,\xi)^2}\right]\left[\frac{\rho(x,\eta_2)}{\rho(x,\xi)}\right][N(x,\lambda)\,\rho(x,\xi)] \qquad \text{qed.}$$

<u>Lemma B.</u>

$$\left|\prod_{j=1}^{\ell}(\nabla_\xi,\eta_j)\prod_{k=1}^{m}(\nabla_x,\lambda_k)(|\xi|^2)\right|$$

$$\leq \rho(x,\xi)^4\prod_j\left[\frac{\rho(x,\eta_j)}{\rho(x,\xi)} + \frac{\rho(x,\eta_j)^2}{\rho(x,\xi)^2}\right]\prod_k\,[\rho(x,\xi)\,N(x,\lambda) + \rho(x,\xi)^2 N(x,\lambda_k)^2]$$

<u>Proof.</u> If $m > 0$, there is nothing to prove. If $m = 0$, $\ell > 2$ there is

also nothing to prove. Now:

$\ell = 0 \qquad |\xi|^2 \le \rho(x,\xi)^4 .$

$\ell = 1 \qquad |\langle \xi, \eta\rangle| \le |\xi|\,|\eta| \le \rho(x,\xi)^4 \left[\frac{\rho(x,\eta)^2}{\rho(x,\xi)^2}\right].$

$\ell = 2 \qquad |\langle \eta_1\, \eta_2\rangle| \le |\eta_1|\,|\eta_2| \le \rho(x,\xi)^4 \prod_{j=1}^{2} \left[\frac{\rho(x,\eta_j)^2}{\rho(x,\xi)^2}\right]$ qed.

Proposition. (The class $\tilde{S}^m_\rho$)

(a) $a_1, a_2 \in \tilde{S}^m_\rho,\ c_1, c_2 \in \mathbb{C} \implies c_1 a_1 + c_2 a_2 \in \tilde{S}^m_\rho$

(b) $a \in \tilde{S}^{m_1}_\rho,\ b \in \tilde{S}^{m_2}_\rho \implies ab \in \tilde{S}^{m_1+m_2}_\rho .$

(c) $a \in \tilde{S}^m_\rho \implies \frac{\partial a}{\partial \xi_j} \in \tilde{S}^{m-1}_\rho,\ \frac{\partial a}{\partial x_j} \in \tilde{S}^{m+2}_\rho$

(d) $a \in \tilde{S}^m_\rho$ with $a > 0$, and

if $a \approx \rho^m$ then $a^z \in \tilde{S}^{m\,\mathrm{Re}(z)}_\rho$, $\forall z \in \mathbb{C}$.

Proof. (a), (b), (c) are all clear. To check (d), note that

$\prod_{j=1}^{\ell} (\nabla_\xi, \eta_j) \prod_{k=1}^{m} (\nabla_x, \lambda_k)\, (a(x,\xi)^z)$ is a sum of terms of the form:

$$(\text{coeff})\ a(x,\xi)^{z-k}\, ((\nabla_\xi,\eta)^{a_1} (\nabla_x,\lambda)^{b_1} a(x,\xi)) \ldots ((\nabla_\xi,\eta)^{a_k} (\nabla_x \lambda)^{b_k} a(x,\xi))$$

where $\sum a_j = \ell$, $\sum b_j = m$. We estimate each such term in absolute value by

$$\lesssim a(x,\xi)^{\mathrm{Re}(z)-k} \prod_{j=1}^{k} \left[\rho(x,\xi)^m\, [\mathrm{I}]^{a_j}\, [\mathrm{II}]^{b_j}\right]$$

(where [I], [II] are the usual factors)

$$\lesssim \rho(x,\xi)^{mk} a^{Re(z)-k} [I]^{\ell} [II]^{m}$$

$$\lesssim \rho(x,\xi)^{m\,Re(z)} [I]^{\ell} [II]^{m}$$

provided $\quad a^{Re(z)-k} \leq \rho(x,\xi)^{m(Re(z)-k)}$ qed.

Corollary. $\quad \rho(x,\xi)^{z} \in \tilde{S}_{\rho}^{Re(z)}$

Notice that by Lemmas A and B, $\rho^{4} \in \tilde{S}_{\rho}^{4}$.

Corollary. Let $\varphi \in C_0^{\infty}(\mathbb{R})$. Then $\varphi(\rho(x,\xi)) \in \tilde{S}_{\rho}^{0}$.

§3. The class S_{ρ}^{m}

Theorem. (a) S_{ρ}^{m} is a complex vector space

(b) $a_j \in S_{\rho}^{m_j}$, $j=1,2 \implies a_1 a_2 \in S_{\rho}^{m_1+m_2}$

(c) $S_{\rho}^{m} \subset \tilde{S}_{\rho}^{m}$

(d) $a \in S_{\rho}^{m}$, $a > 0$, $a \approx \rho^{m} \implies a^{z} \in S_{\rho}^{m\,Re(z)}$

(e) $Q(x,\xi) \in S_{\rho}^{2}$, $|\xi|^{2} \in S_{\rho}^{4}$

(f) $\rho(x,\xi)^{z} \in S_{\rho}^{Re(z)}$

Proof. (a) and (b) are again obvious. (f) follows from (e) and (d) and (a).

To prove (c), let $a \in S_{\rho}^{m}$. We must estimate

$$|(\nabla_{\xi},\eta)^{j} (\nabla_{x},\lambda)^{k} a(x,\xi)|.$$

But $(\nabla_x \lambda)^k a$ is a sum of terms of the type

$$|\lambda|^p \; \langle \nabla \varphi(x), \lambda \rangle \;\; Q^1(x, \xi)^q \;\; b_p(x, \xi, \lambda) \;\; c_{m-p-2q}(x, \xi)$$

where $p + q = k$. Hence we have to estimate terms of the following type:

$$|\lambda|^p \; \langle \nabla \varphi(x), \lambda \rangle^q \; (\nabla_\xi, \eta)^{j_1} (Q^1(x, \xi))^q \; (\nabla_\xi, \eta)^{j_2} b(x, \xi, \lambda) \, (\nabla_\xi, \eta)^{j_3} c$$

where $j_1 + j_2 + j_3 = j$.

Now $$|(\nabla_\xi, \eta)^{j_3} \; c(x, \xi)| \leq \rho(x, \xi)^{m-p-2q} \, [I]^{j_3}$$

$$|(\nabla_\xi \, \eta)^{j_2} \; b(x, \xi, \lambda)| \lesssim |\eta|^{j_2} \, |\xi|^{p-j_2}, \quad \text{and}$$

$$(\nabla_\xi, \eta)^{j_1} \, (Q^1(x, \xi)^q) = \sum_{\ell \leq q} \; Q(x, \xi)^{q-\ell} \, ((\nabla_\xi, \eta)^{\ell_1} \, Q^1) \ldots ((\nabla_\xi, \eta)^{\ell_1} \, Q^1)$$

where $\sum \ell_\alpha = j_1$.

Here a typical term is bounded by

$$Q^1(x, \xi)^{q-\ell} \, (Q^1(x, \xi)^{1/2})^B \;\; (Q^1(x, \eta)^{1/2})^{\ell + A}$$

where $B = \#$ of $\ell_j = 1$

and $A = \#$ of $\ell_j = 2$

(Then $A + B = \ell$, and $\ell + A = j_1$.)

Therefore, $$|(\nabla_\xi, \eta)^{j_1} \, (Q^q)| \lesssim Q^1(x, \xi)^{q-\ell+B/2} \; Q^1(x, \eta)^{(\ell + A)/2}.$$

Finally, $$|(d\varphi, \lambda)|^q \lesssim N_x(\lambda)^q \varphi(x)^q + N_x(\lambda)^{2q}.$$

Hence, one term we have to estimate is:

$$|\lambda|^p N_x(\lambda)^q \varphi(x)^q Q^1(x,\xi)^{q-\ell+B/2} Q^1(x,\eta)^{\ell+A/2} |\eta|^{j_2} |\xi|^{p-j_2} \rho(\xi)^{m-p-2q} [I]^{j_3}$$

$$\lesssim \rho(\xi)^m N_x(\lambda)^{p+q} \rho(\xi)^{2q-2\ell+B} \rho(\eta)^{\ell+A} |\eta|^{j_2} |\xi|^{p-j_2} \rho(\xi|^{-p-29} [I]^{j_3}$$

$$= \rho(\xi)^m [N_x(\lambda)\rho(\xi)]^{p+q} \rho(\xi)^{-p-q} \rho(\xi)^{2q} [I]^{j_1} [I]^{j_3} |\eta|^{j_2} |\xi|^{j_2} |\xi|^{p-j_2} \rho(\xi)^{-p}$$

$$\lesssim \rho|\xi|^m [II]^k [I]^{j_1+j_3} \rho(\eta)^{2j_2} |\xi|^{p-j_2} \rho(\xi)^{-2p-q}$$

$$= \rho(\xi)^m [I]^{j_1+j_2+j_3} [II]^k |\xi|^{p-j_2} \rho(\xi)^{-2p-q+2j_2}$$

$$\lesssim \rho(\xi)^m [I]^{j_1+j_2+j_3} [II]^k,$$

since $p-j_2 \geq 0$, and therefore

$$|\xi|^{p-j_2} \leq \rho(\xi)^{2p-2j_2}, \quad \text{so}$$

$$|\xi|^{p-j_2} \rho(\xi)^{-2p+2j_2-q} \leq \rho(\xi)^{-q} \lesssim 1.$$

The other term we have to estimate is:

$$|\lambda|^p N_x(\lambda)^{2q} Q^1(x,\xi)^{q-\ell+B/2} Q^1(x,\eta)^{\ell+A/2} |\eta|^{j_2} |\xi|^{p-j_2} \rho(\xi)^{m-p-2q} [I]^{j_3}$$

$$\lesssim \rho(\xi)^m [II]^k \rho(\xi)^{-p-2q} |\xi|^{2q-2\ell+B} |\eta|^{\ell+A} |\eta|^{j_2} |\xi|^{p-j_2} \rho(\xi)^{-p-2q} [I]^{j_3}$$

$$= p(\xi)^m [II]^k \rho(\xi)^{-p-2q} |\xi|^{2q-j_1} |\eta|^{j_1} |\eta|^{j_2} |\xi|^{p-j_2} \rho(\xi)^{-p-2q} [I]^{j_3}$$

$$= \rho(\xi)^m [II]^k [I]^{j_1+j_2+j_3} \rho(\xi)^{2j_1} \rho(\xi)^{2j_2} p(\xi)^{-2p-4q} |\xi|^{2q-j_1} |\xi|^{p-j_2}$$

$$\lesssim \rho(\xi)^m [I]^j [II]^k \left[|\xi|^{2q-j_1} p(\xi)^{-4q+2j_1} \right] \left[|\xi|^{p-j_2} \rho(\xi)^{-2p+2j_1} \right]$$

$$\leq \rho(\xi)^m\,[I]^j\,[II]^k, \quad \text{since } j_1 \leq 2q \quad j_2 \leq p.$$

This proves (c).

To prove (d), let $a \in S^m_\rho$. Then $a \in \tilde{S}^m_\rho$ so $a^z \in \tilde{S}^{m\,Re(z)}_\rho \subset \hat{S}^{m\,Re(z)}_\rho$ (by the proposition). To show that $a^z \in S^{m\,Re(z)}_\rho$, we shall have to consider the behavior of x derivatives of a^z (see §7). Let us consider in detail $(\nabla_x, \lambda)^j (a^z)$. A typical term resulting from the differentiation is a multiple of

$$a^{z-k}(x,\xi)\,((\nabla_x,\lambda)^{b_1} a)\ldots((\nabla_x,\lambda)^{b_n} a)$$

where $\sum b_\ell = j$.

Since $a \in S^m_\rho$ each derivative occurring above is a sum of terms; (see the identity following (2) in §7). This leads us to terms of the form

$$a^{z-k}(x,\xi) \prod_{\ell=1}^{k} \left\{ |\lambda|^{p_\ell}\, |(d\varphi,\lambda)|^{q_\ell}\, Q_1(x,\xi)^{q_\ell}\, b_{p_\ell}(x,\xi,\lambda)\, c_{m-p_\ell-2q_\ell}(x,\xi) \right\}$$

where $p_\ell + q_\ell = b_\ell$.

Now $\prod b_\ell$ is a polynomial of degree $\sum p_\ell$, while $\prod c_{m-p_\ell-2q_\ell} \in \hat{S}_\rho^{km-\sum p_\ell - 2\sum q_\ell}$. Also $a \in S^p_m \subset \tilde{S}^p_m$. Therefore by the proposition $a^{z-k} \in \tilde{S}^{m(Re(z)-k)} \subset \hat{S}_\rho^{m(Re(z)-k}$, and hence

$$a^{z-k} \prod c_{m-p_\ell-2q_\ell} \in \hat{S}^N_\rho, \quad \text{where } N = m\cdot Re(z) - \sum p_\ell - 2\sum q_\ell$$

which is the desired inclusion for the proof of (d).

Finally, to prove (e) note that for $j \geq 2$ $(\nabla_x,\lambda)^j Q$ is a quadratic polynomial (in ξ), so satisfies the requirements. On the other hand

$$(\nabla_x, \lambda)Q = \varphi(x)\,(\nabla_x Q^1(x,\xi), \lambda) + (\nabla\varphi, \lambda)\, Q^1(x,\xi)$$

Now $Q^1(x,\xi) = \Sigma\,|L_j(x,\xi)|^2$, so it remains to check that $\varphi(x)\,L_j(x,\xi) \in \hat{S}^1_\rho$. But $(\nabla_\xi, \eta)\,(\varphi(x)\,L_j(x,\xi)) = \varphi(x)\,L_j(x,\eta) \leq \rho(x,\xi)\left[\frac{\rho(x,\eta)}{\rho(x,\xi)}\right]$, so this holds. qed.

References

1. R. Beals, "A general calculus of pseudo-differential operators," Duke Math. J. (1975) 42, 1-42.

2. ——————, "L^p and Schauder estimates for pseudo-differential operators," to appear.

3. R. Beals and C. Fefferman, "Spatially inhomogeneous pseudo-differential operators," Comm. Pure Appl. Math (1974) 27, 1-24.

4. ——————, "On the hypoellipticity of second-order operators," Comm. Partial Diff. Equations (1976) 1, 73-85.

5. L. Boutet de Monvel, "Hypoelliptic operators with double characteristics and related pseudo-differential operators," Comm. Pure Appl. Math (1974) 27, 585-639.

6. L. Boutet de Monvel and J. Sjöstrand, "Sur la singularité des noyaux de Bergman et de Szegö," Astérisque (1976) 34-35, 123-164.

7. L. Boutet de Monvel and F. Treves, "On a class of pseudo-differential operators with double characteristics," Inventiones Math (1974) 24, 1-34.

8. A. P. Calderón, "Lebesgue spaces of differentiable functions and distributions," Amer. Math. Soc. Proc. Symp. Pure Math 5(1961), 33-49.

9. A. P. Calderón and R. Vaillancourt, "A class of bounded pseudo-differential operators," Proc. Nat. Acad. Sci. (1972) 79, 1185-1187.

10. R. R. Coifman and G. Weiss, "Analyse harmonique non-communicative sur certains éspaces homogènes," Lecture Notes in Mathematics (1971) no 242, Springer Verlag.

11. V. Yu, Egorov and V. A. Kondrater, "The oblique derivative problem," Math. USSR Sbornik (1969) 7, 139-169.

12 E. B. Fabes and N. M. Rivière, "Singular integrals with mixed homogeneity," Studia Math. (1966) 27, 19-38.

13. C. Fefferman, "The Bergman kernel and biholomorphic mappings of pseudo-convex domains," Invent. Math. (1974) 26, 1-66.

14. G. B. Folland, "Subelliptic estimates and function spaces on nilpotent Lie groups," Arkiv f. Mat. (1975) 13, 161-207.

15. G. B. Folland and J. J. Kohn, "The Neumann problem for the Cauchy-Riemann complex," Annals of Math. Studies (1972) no. 75, Princeton University Press.

16. G. B. Folland and E. M. Stein, "Estimates for the $\bar{\partial}_b$ complex and analysis on the Heisenberg group," Comm. Pure and Appl. Math (1974) 27, 429-522.

17. L. Gärding, Bulletin Soc. Math. France (1961) 89, 381-428.

18. R. Goodman, "Nilpotent Lie groups," Lecture Notes in Mathematics (1976) no 562, Springer Verlag.

19. P. C. Greiner, J. J. Kohn, and E. M. Stein, "Necessary and sufficient conditions for solvability of the Lewy equation," Proc. Nat. Acad. Sci. (1975) 72, 3287-3289.

20. P. C. Greiner and E. M. Stein, " Estimates for the $\bar{\partial}$-Neumann problem," Mathematical Notes (1977) no 19, Princeton University Press.

20a. V. V. Grushin, "On a class of hypoelliptic pseudo-differential operators degenerate on a sub-manifold," Math. USSR Sbornik (1971) 13, 155-185.

21. S. Helgason, "Differential geometry and symmetric spaces," (1962) Academic Press, New York.

22. I. I. Hirschman Jr., "Multiplier transformations I," Duke Math. Jour. (1956) 26, 222-242.

23. L. Hörmander, "Pseudo-differential operators and non-elliptic boundary problems," Ann. Math. (1966) 83, 129-209.

23a. ——————, "Hypoelliptic second-order differential equations," Acta Math. (1967) 119, 147-171.

24. ——————, "Pseudo-differential operators and hypoelliptic equations," Amer. Math. Soc. Proc. Symp. Pure Math. (1967) no. 10, 138-183.

25. ——————, "The Weyl calculus of pseudo-differential operators," to appear.

26. Y. Kannai, "An unsolvable hypoelliptic differential operator," Israel J. Math. (1971) 9, 306-315.

27. N. Kerzman and E. M. Stein, "The Szegö kernel in terms of Cauchy-Fantappiè kernels," Duke Math. Jour. (1978) 45, 197-224.

28. A. W. Knapp and E. M. Stein, "Intertwining operators for semi-simple groups," Ann. of Math. (1971) 93, 489-578.

29. A. Korányi and S. Vagi, "Singular integrals in homogeneous spaces and some problems of classical analysis," Ann. Scuola Norm. Sup. Pisa (1971) 25, 575-648.

30. S. Krantz, "Generalized function spaces of Campanato type," to appear.

31. P. Kree, "Distributions quasi-homogènes," C.R.A. Sci. Paris (1965) 261, 2560.

32. J. L. Lions and J. Peetre, "Sur une classe d'espaces d'interpolation," Publ, Math. Inst. Hautes Etudes Sci. (1964) 19, 5-68.

33. W. Madych and N. Rivière, "Multipliers of Hölder classes," Jour. of Funct. Analysis (1976) 21, 369-379.

34. A. Nagel and E. M. Stein, "A new class of pseudo-differential operators," Proc. Nat. Acad. Sci (1978) 75, 582-585.

35. ——————, "Some new classes of pseudo-differential operators," Proc. Symp. Amer. Math. Soc. held in Williamstown, Summer 1978, to appear.

36. R. O'Neil, "Two elementary theorems on the interpolation of linear operators," Proc. Amer. Math. Soc. (1966) 17, 76-82.

37. D. H. Phong and E. M. Stein, "Estimates for the Bergman and Szegö projections on strongly pseudo-convex domains," Duke Math. Jour. (1977) 44, 695-704.

38. N. M. Rivière, "Singular integrals and multiplier operators," Arkiv f. Mat. (1971) 9, 243-278.

39. L. P. Rothschild and E. M. Stein, "Hypoelliptic differential operators and nilpotent groups," Acta Math. (1976) 137, 247-320.

40. J. Sjöstrand, "Operators of principal type with interior boundary conditions," Acta Math. (1973) 130, 1-51.

41. E. M. Stein, "Singular integrals and differentiability properties of functions," (1970), Princeton University Press.

42. ——————, "Singular integrals and estimates for the Cauchy-Riemann equations," Bull. Amer. Math. Soc. (1973) 79, 440-445.

43. M. H. Taibleson, "Translation invariant operators, duality, and interpolation II," J. Math. Mech. (1965) 14, 821-840.

44. S. Wainger, "Special trigonometric series in K dimensions," Mem. Amer. Math. Soc. (1965) no 59.

45. N. Wiener, "The Fourier integral and certain of its applications, (1933), Cambridge Univ. Press.

Library of Congress Cataloging in Publication Data

Nagel, Alexander, 1945-
Lectures on pseudo-differential operators.

Includes bibliographical references.
1. Pseudodifferential operators.
I. Stein, Elias M., 1931- joint author.
II. Title.
QA329.7.N34 515'.72 79-19388
ISBN 0-691-08247-2

Milton Keynes UK
Ingram Content Group UK Ltd.
UKHW020720101124
450942UK00006B/294

9 780691 601090